MODELING AND SIMULATION FOR RF SYSTEM DESIGN

Modeling and Simulation for RF System Design

by

RONNY FREVERT
Fraunhofer Institute for Integrated Circuits, Dresden, Germany

JOACHIM HAASE
Fraunhofer Institute for Integrated Circuits, Dresden, Germany

ROLAND JANCKE
Fraunhofer Institute for Integrated Circuits, Dresden, Germany

UWE KNÖCHEL
Fraunhofer Institute for Integrated Circuits, Dresden, Germany

PETER SCHWARZ
Fraunhofer Institute for Integrated Circuits, Dresden, Germany

RALF KAKEROW
Nokia Research Center, Bochum, Germany

and

MOHSEN DARIANIAN
Nokia Research Center, Bochum, Germany

Springer

A C.I.P. Catalogue record for this book is available from the Library of Congress.

ISBN 10 0-387-27584-3 (HB)
ISBN 13 978-0-387-27584-0 (HB)
ISBN 10 0-387-27585-1 (e-book)
ISBN 13 978-0-387-27585-7 (e-book)

Published by Springer,
P.O. Box 17, 3300 AA Dordrecht, The Netherlands.

www.springeronline.com

The book and the included CD-ROM contain models which may be used for simulation purposes. The user accepts full responsibility for the use of these models.

The names of software products used in this book are trademarks of their respective producers.

Printed on acid-free paper

Printed in the Netherlands.

Contents

Preface

Many books have been published in recent years that focus on wireless communication systems, with some focused on modeling and simulation. This book is aimed at the special topic of modeling for RF system design. Very high carrier frequencies together with long observation periods result in extremely large computation times and requires, therefore, specialized modeling methods and simulation tools on all design levels from system down to circuit level. To illustrate the application of these methods and usage of the tools the book includes numerous models and extensive examples. Therefore the book is addressed to graduate students and industrial professionals who are engaged in communication system design and want to gain insight into the system structure by own simulation experiences.

The tools and languages for hardware description of VLSI circuits have changed over the years. Nevertheless models are provided on a CD-ROM included with this book because models are necessary to reproduce, understand and explore the real world behavior on a simulation platform. VHDL-AMS and Verilog-A are chosen as description languages which are an IEEE standard and a quasi industrial standard respectively. In spite of deviations within language implementations in different simulation tools, the provided mathematical background to each individual model should enable a large audience of readers to use these models. Moreover the given introduction into the syntactic elements of the language VHDL-AMS allows to modify the given examples to special needs.

The authors

Acknowledgments

This book is the result of many years of fruitful project cooperation between Nokia Research Center, the Fraunhofer Institute for Integrated Circuits and other partners. After common discussions and successful research in the field of modeling methodology for wireless system design we were convinced that it is time to publish our approaches, methods and results together with illustrating examples.

The authors are grateful to all colleagues inside and outside of our organizations for sharing their knowledge during discussions and to all supporters who helped with their valuable hints and corrections to complete the work on this book. Especially we wish to thank Dean Hobson from Mentor Graphics who carefully read the manuscript and was always prepared to discuss matters of language and content. Also, we would like to thank Mark de Jongh for his encouraging hints and the management task to publish this book. This also includes of course the staff at Kluwer publishers who produced this book in a very professional way.

Chapter 1

INTRODUCTION

Modern telecommunication systems are highly complex from an algorithmic point of view. The complexity continues to increase due to advanced modulation schemes, multiple protocols and standards, as well as additional functionality such as color displays, personal organizers, navigation aids, cameras, and audio-visual support.

At the same time both silicon area – which means costs – and power consumption of the devices have to be reduced and the design time shortened. This is inevitable to keep profitability in this fast evolving high volume consumer market.

These conflictive demands force the need for efficient design and verification methods. To have short and reliable design cycles, verification is necessary very early in the design process. Modeling and simulation need to accompany the design steps from the specification to the overall system verification in order to bridge the gaps between system specification, system simulation, and circuit level simulation. Therefore this book contains application-oriented training material for RF designers which combines the presentation of a mixed-signal design flow, an introduction into the standardized powerful hardware description language VHDL-AMS, and the application of commercially available simulators. The focus lies on RF specific modeling and simulation methods and the consideration of system and circuit level descriptions.

An early version of some parts of this book, especially some of the VHDL-AMS models, has been tested in a Nokia-internal course with about 50 designers. In this course a web-based education and simulation environment has been used, developed in a European research project LIMA (Learning Platform in Microelectronics Applications).

The challenges for the designer are especially demanding in the face of mixed-signal (analog/digital) and multi-domain (RF/baseband) systems. Today's wireless communication systems use sophisticated modulation and coding techniques to transmit the information at very high carrier frequencies. Modulation and coding is typically realized in the Digital Signal Processing (DSP) subsystem, which is also called baseband signal processing. The RF front-end provides the interface between baseband (some MHz) and the RF transmission channel (some GHz).

The DSP part uses more than 95% of the total amount of transistors. System level simulators are used for the verification of the DSP algorithms. Efficient simulation algorithms are applied to simulate the complete transmit path from the transmitter to the receiver. DSP designers often assume that the analog part is an ideal device. On the other hand RF designers perform analog simulations to design and verify the RF subsystem without information regarding the DSP part. This is why the common evaluation of the RF and the DSP part becomes increasingly important. This ensures that the RF part fulfills the system requirements without over-dimension, which means the interaction between both parts is respected without the need to include a safety margin in the specification of the RF part.

RF circuits and systems possess special characteristics that need to be considered in modeling and simulation, which are

- very high carrier frequency on the one hand and comparatively low signal bandwidth on the other,
- presence of weak nonlinearities,
- importance of noise considerations and the signal-to-noise ratio (SNR),
- necessity to simulate a large number of sample points or data bits in order to compute distortion measures, for example bit error rates (BER).

For RF systems to handle these characteristics specially suited modeling methods and simulation algorithms have been developed. They will be introduced during the course of this book and demonstrated with examples.

A number of simulation tools are on the market that specialize in RF circuits. Since we want to widen the scope on a design flow from system to circuit level with attention to mixed-signal aspects, we used a collection of different commercially available simulation tools in the book

- ADVance MS of Mentor Graphics
- SpectreRF of Cadence
- SPW of CoWare
- MATLAB of The MathWorks

Many other tools currently available on the market could have been used, but the modeling methods and simulation principles remain the same. An introduction into the usage of the tools goes beyond the scope of this book. For support on the tools, refer to the help function or the online help of the tool providers. It is also not intended to include schematic entry and layout tools.

Modeling of RF systems ranges from system-level signal-flow oriented models (for example MATLAB/Simulink) over mixed-signal block oriented models (for example VHDL-AMS) to circuit-level descriptions (for example SpectreRF). Therefore a modeling flow, covering different levels of abstraction, as well as modeling languages and libraries are essential topics of the book (Figure 1-1). A special focus lies on the mixed-signal simulator-independent modeling language VHDL-AMS.

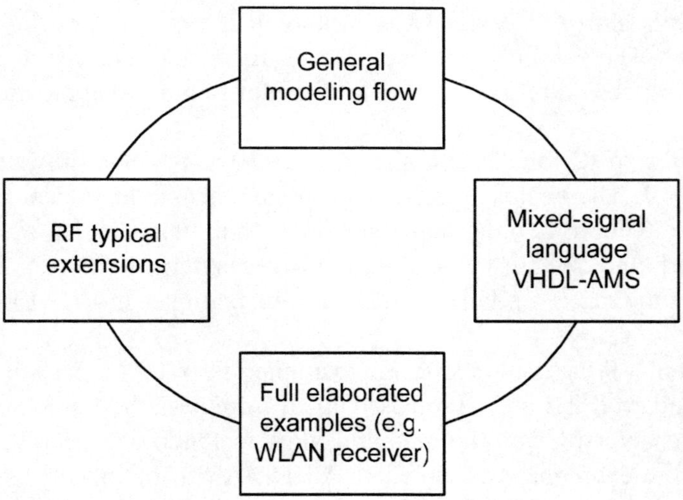

Figure 1-1. Overview of the main topics of the book

Modeling and simulation methods need to be oriented on existing design flows in order to establish them in industrial use. Hence we propose a modeling and simulation flow that follows the V-diagram as a commonly accepted design paradigm (see Chapter 2). The material in this book is structured accordingly. Chapter 2 provides an overview of different levels of abstraction, the top-down and bottom-up methodologies. Specific simulation algorithms and various simulation tools for different phases of RF system design are introduced in Chapter 3.

The first direction of the design flow is top-down. That means we start with specifications at the system level. Chapter 4 describes how RF components can be modeled in system level simulators such as CoCentric, SPW or MATLAB. It is focused on the development of RF-specific system models.

After initial architectural decisions, specifications for the subsystems are derived and an abstract (less detailed) behavioral model of the RF subsystem can be developed for simulation. This model is improved and becomes more detailed during the design process. On this architecture or block level, mixed-signal simulations are often necessary because the partition into analog and digital parts is not yet clear and different architectures have to be explored. At this point in the book we introduce VHDL-AMS as an important language that supports digital, analog, and mixed-signal modeling and simulation. It is a strict superset of the digital VHDL 1076-1993. Chapter 6 is aimed at designers with knowledge of standard digital VHDL 1076-1993. The reader should be able to understand and use the provided models, change and refine them, as well as develop own simple models.

A library of RF block level models in VHDL-AMS is fully documented in Chapter 7. The enclosed CD-ROM contains the complete source code of this model library. Important basic RF building blocks are included subdivided into source, processing and measurement blocks. Chapter 8 introduces the macromodeling principle with examples in VHDL-AMS.

In Chapter 9 the complex design example of a WLAN receiver according to the standard IEEE 802.11a is assembled from basic building blocks of the previous chapters. Using the modeling flow methodology from the previous chapters the example is modeled in VHDL-AMS, optimized using circuit level simulation, and verified by system level simulation. Thereby it is shown, how the realistic design task of developing a receiver front-end can be supported by modeling and simulation.

The next step in the top-down design flow is the implementation of blocks as circuits. At this level, circuit simulators are available with dedicated support for RF analysis and depiction modes. The custom IC design environment from Cadence and its analog RF simulator SpectreRF are important tools in RF circuit design. SpectreRF uses Verilog-A for behavioral modeling, which is the analog part of Verilog-AMS. A library of Verilog-A models for typical RF building blocks is provided by Cadence. Chapter 10 demonstrates the use of this library for RF system modeling. An example of modeling in Verilog-A is provided.

Bottom-up techniques are used next in the design flow to verify whether design goals are met with the implemented system. The characterization of circuit level descriptions allows the refinement of behavioral models for system level simulation. It is also applied to generate data for the component documentation and reuse. Characterization environments are discussed in the Chapter 11. The characterization environment is used to extract RF specific parameters of circuit designs and to validate the respective behavioral models. An overview of parameters, which can be extracted for RF components, is provided. A characterization example is demonstrated by using SpectreRF and OCEAN scripts.

As a last step in the design flow, system verification is necessary with the back-annotated knowledge of the circuit properties in the refined models. Solutions which will bring analog and system level simulators together are introduced in the last Chapter 12. Black box modeling uses a special kind of characterization to generate nonlinear transfer functions of a complete RF front-end. The transfer functions are stored in files which are read from special black box models in the system level simulator. Another method is co-simulation, which couples analog and system level simulators. The principles of both approaches are explained and illustrated by examples for the Cadence design environment. Advantages and disadvantages of the different approaches are discussed.

To summarize, the training material comprises up-to-date knowledge of modeling and simulation for the RF system design of modern telecommunication systems. The introduction of a general modeling flow is supplemented by RF specific simulation algorithms. Commercially available tools are used to demonstrate how RF system design can be supported and improved by means of modeling and simulation. A second major part is the introduction of VHDL-AMS as a standardized hardware description language with increasing importance. Because it is the mixed-signal extension of the well-established language VHDL it is expected to be used for RF and system design tasks in the near future.

In this application-oriented book the teaching material, which introduces the concepts and theoretical background, is followed by illustrative examples and sources of further information. Many simulation examples are shown with extensive solutions. Thus if the reader has access to the required simulation tools he is able to reproduce the example solution, modify it and thereby gain own experiences with modeling and simulation of RF systems. This book establishes a comprehensive training course in a technologically critical area.

Chapter 2

DESIGN FLOW OVERVIEW

2.1 Design Levels

Functionality and architecture of electronic devices can be very complex. The systems may consist of analog and digital hardware together with software parts. A telecommunication system contains for example:

- An analog front-end to the physical transmission channel
- Digital hardware for coding and modulation
- General purpose or signal processors for control, user interface and transmission protocol handling

Many designers with specialization in different areas are involved in design and implementation. Several design steps are necessary to realize a system concept on silicon. The design process can be classified in several design levels as shown in Figure 2.1.

Each design level is associated with certain design tasks concerning the whole system or system parts. Starting from system level the design description becomes more and more detailed in a design step. CAD tools support the designer at each level.

The system level is the first design level beginning with an idea of the desired system. This level is also called concept engineering. The system concept and main algorithms are described at a very abstract level without information about the implementation of algorithms. For example, the coding algorithm to be used for data transmission is specified, but it is not decided to implement the coder in hardware or software.

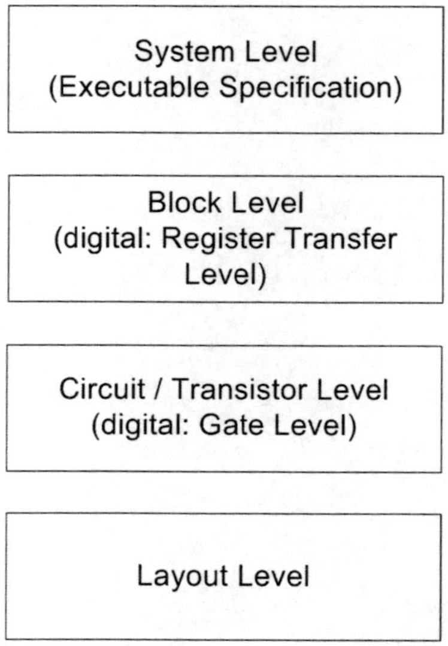

Figure 2-1. Design levels

The system specification can be developed on a sheet of paper. More powerful is an executable specification supported by system-level simulators (for example CoCentric System Studio, MATLAB, and SPW). It allows the evaluation of the selected algorithms and provides a reference model for following design steps.

The system is now partitioned into several hardware (analog or digital) and software subsystems. This design level is named Block Level or Register Transfer Level (RTL) in the digital area. The description of the subsystems at this level contains more detail about the design architecture. At this level the design consists of different blocks, for example multiplier, adder, register, A/D converter, analog filter and amplifier.

Digital and mixed-signal hardware description language (HDL) simulators support the block level design. Commonly used modeling languages in this area are VHDL-AMS and Verilog-AMS. The design of hardware/software systems is further supported by special tools, for example instruction set simulators (ISS).

The third design level is called gate level in the digital domain and circuit level in the analog domain. The blocks of the system are now represented by netlists containing gates or active and passive analog elements. Gate level models can be generated from RTL descriptions by logic synthesis. In the analog design, the circuits are still designed manually.

Gate level or circuit simulation is used to evaluate the design at block level. In the digital domain a timing analysis can be executed, and the blocks are still described in VHDL and Verilog. Circuit simulators such as SPICE and Spectre are used in the analog domain to analyze the behavior of the designed block.

Based on the gate level or circuit netlist and data of the circuit technology the layout of the circuit is designed. The design is now represented as polygons at different layers of an integrated circuit. In the digital domain this step is well-automated. The tools will check if the design rules for a specified circuit technology are fulfilled. In the analog domain further manual optimization of layout may be necessary, for example to minimize crosstalk between signals or to achieve a symmetric design. Tools that extract parasitic effects that originate from layout also support the layout verification.

2.2 Top-down System Design

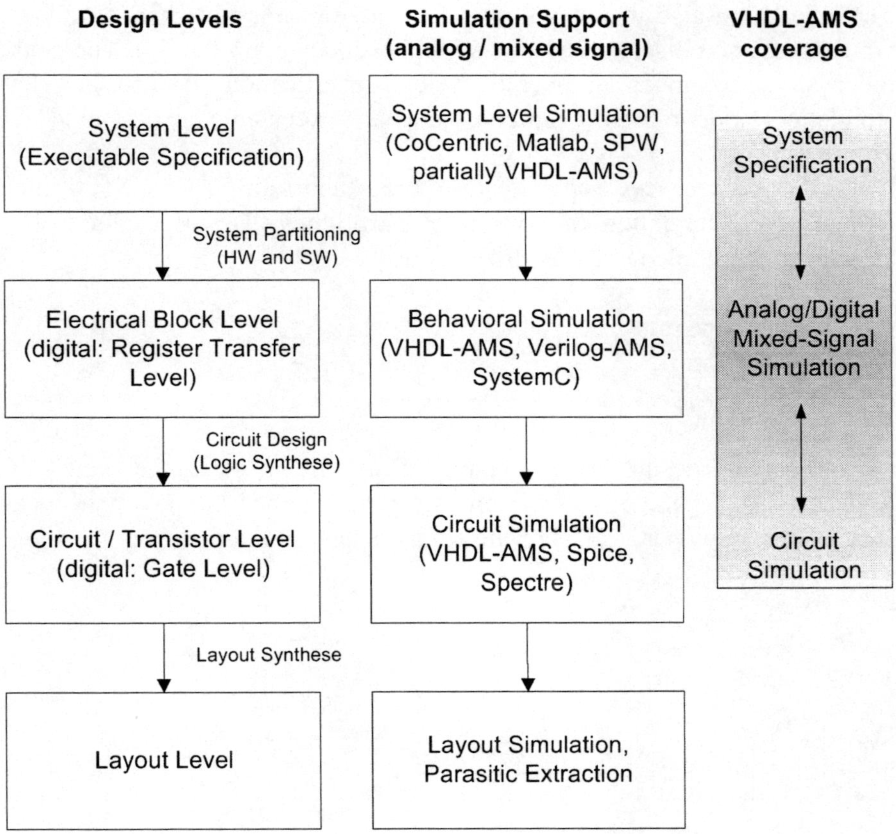

Figure 2-2. Top-down design and simulation support

Top-down design is a method of designing an electronic system that starts with the complete system concept and then breaks it down into smaller and smaller components (see Figure 2-2).

The first design level at which top down design starts is the system level. For telecommunication systems it is here that is specified which algorithms are used to transmit data from the signal source at point A to a sink at point B. Algorithms which are specified at this level may be for example:

- data structure and protocol
- forward error correction techniques (FEC)
- modulation techniques (QPSK, QAM, GMSK, OFDM)
- channel equalization and synchronization

The system level design is supported by system level simulation. Efficient simulation techniques (for example event driven or data stream driven simulation) allow the simulation of the complete transmission system. The simulation also includes a model of the transmission channel (additive white Gaussian noise, AWGN, or mobile channels with fading). The goal of the system design is an overall system specification. If a system level simulation model exists, it can be used as an "executable specification" (see Figure 2-3).

If the system level specification was successfully verified within a system level simulation the system is partitioned. The algorithms of the system can be implemented in different ways:

- analog hardware
- digital hardware
- software

The second design level is named Block Level or in the digital area Register Transfer Level. The system is now partitioned into components and subsystems. Now parameters of the components can be specified.

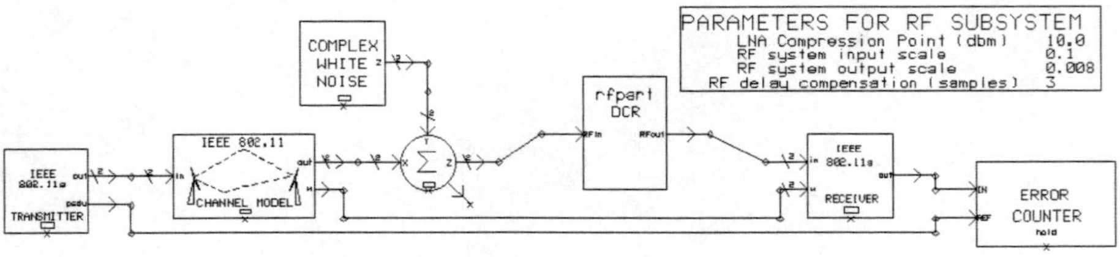

Figure 2-3. Top level schematic of a WLAN system simulation model (SPW)

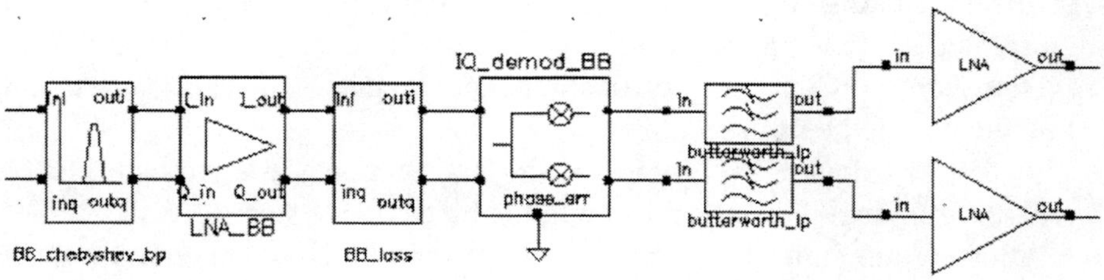

Figure 2-4. Schematic of the RF subsystem (direct conversion receiver)

Figure 2-4 shows for example the block level schematic of the RF subsystem of the WLAN receiver. At system level the RF subsystem was specified either with ideal parameters or with parameters like noise level, gain and linearity. Now it is broken down into its components (filter, amplifier and mixers) which must be parameterized.

At block level we use behavioral models for the simulation of the subsystems. For the analog and mixed-signal area, models can be written in VHDL-AMS and Verilog-AMS. For pure analog simulation, additional languages (for example SpectreHDL) are provided with the simulation tools. The simulation at block level is used to verify whether the block level realization of the subsystem meets the system level requirements.

After the blocks are specified, the circuit design can start. In the digital area, gate level designs can be generated automatically from behavioral models. However for analog blocks there are still no synthesis tools available. So the analog designers must create the transistor level implementation of the components manually. This is supported by transistor level simulation. The block level simulation models can be reused as testbench or reference models if the circuit level simulator supports behavioral modeling languages. Verilog-AMS and VHDL-AMS simulators often support the simulation of SPICE netlists; therefore they can also be used for verification of the transistor level design.

If the transistor level design was verified by simulation the layout can be developed. With the layout level the top down design flow is finished. The layout design is not within the scope of this book. It is possible to extract parasitic effects from layout level simulation which can be used to improve the accuracy of transistor level simulation.

2.3 Bottom-up Verification

The amount of information and number of parameters increases during the top-down design process from the system concept to its implementation.

At the beginning of the design, the system is described with some algorithms. After implementation the system may consist of a large number of transistors. Concept verification is needed to check that the implementation meets the requirements of the system.

In the "V" diagram (Figure 2-5) the verification starts from the layout level (bottom) and then proceeds up to the block and system levels.

After layout, simulation parasitic effects can be back-annotated into the circuit netlist. The circuit simulation with the extracted netlist is used to verify the circuit design. The designed circuits can now be combined into functional blocks, which are checked against their specification in a block level simulation. Finally the designed blocks can be connected to the system. System level simulation verifies that the blocks fit into the system environment.

It is recommended to start verification before the design is completed at layout level. After each design step simulation can be used to verify the design or component against the specification.

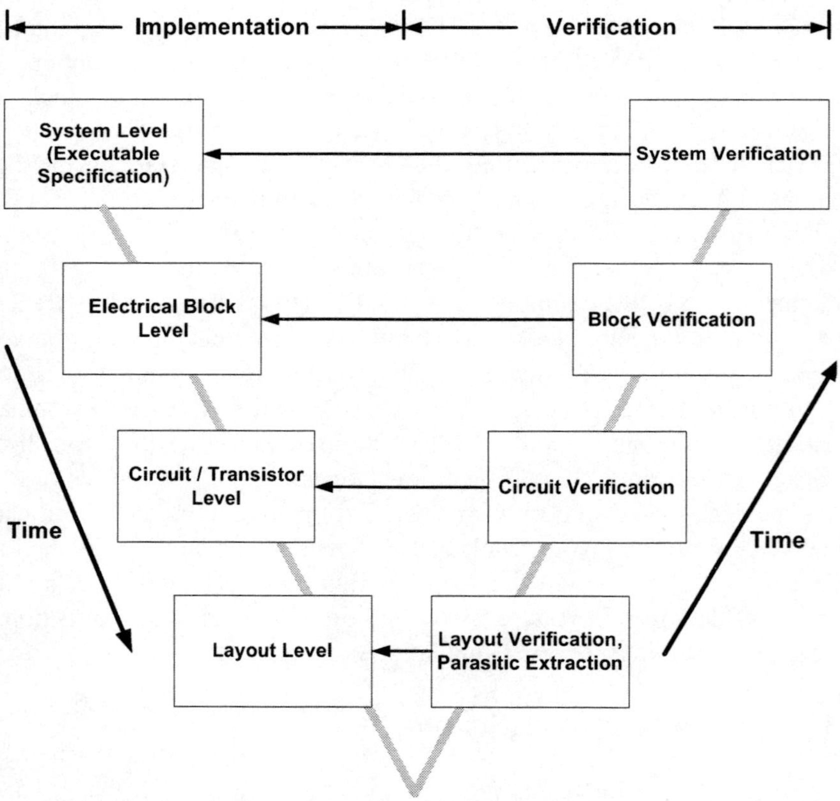

Figure 2-5. Top -down design and bottom-up verification (V diagram)

System level or block level simulation is used to verify large systems or circuits. Often a transistor level model of a system cannot be simulated because its complexity (number of transistors or gates) is much too large. Therefore it is necessary to use behavioral models.

Figure 2-6 shows the application of behavioral models during block level and system level verification. It is assumed that behavioral models were already used during the top-down design. In the verification phase it is now necessary to calibrate these models as follows:

- Parasitic extraction and back annotation into the circuit netlist improves the accuracy of the circuit model (extracted circuit model)
- Simulation with the extracted circuit model is used to gain the circuit characteristic and parameters
- Extracted circuit parameters are used to calibrate the behavioral model of this component
- Calibrated behavioral models are used on block and system levels for verification

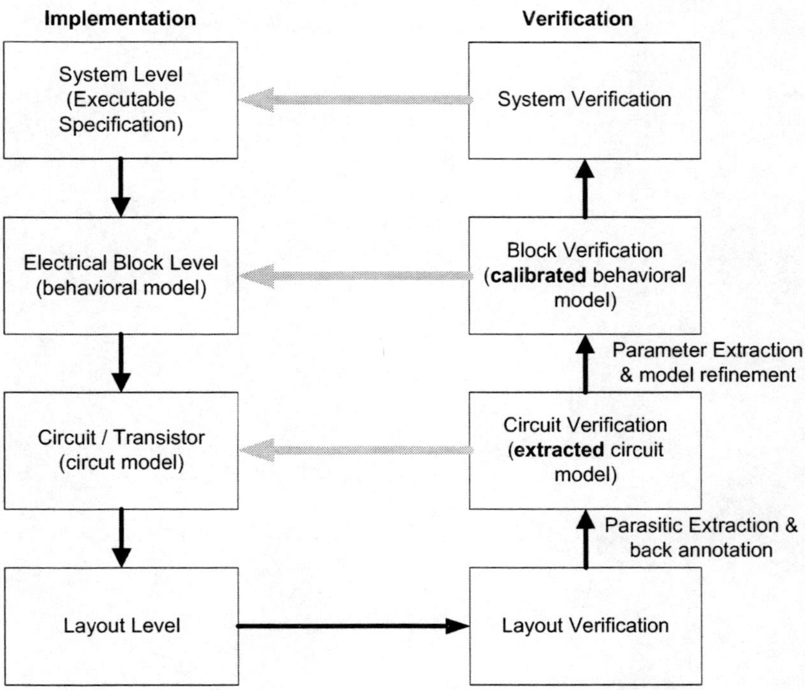

Figure 2-6. Refinement of models during bottom-up verification

The main advantage of using (calibrated) behavioral models is the simulation speedup which enables the simulation of large systems or subsystems.

Different behavioral modeling languages exist. Most of them are specific to a particular simulator. To allow the reuse of models it is suggested to use standardized languages like VHDL-AMS and Verilog-AMS.

A characterization environment can support model calibration. Characterization is the calculation of component or subsystem characteristics and parameters from measured or simulated data. A characterization run contains a set of simulation and postprocessing commands that allow the determination of significant circuit characteristics. The behavior of the circuit description and behavioral model can be compared. If the model is inaccurate, the model parameters or algorithms are modified. Characterization also supports model and circuit documentation. Chapter 11 contains more information about characterization environments.

Chapter 3

SIMULATION TOOLS IN SYSTEM DESIGN

3.1 Use of Simulation Tools within the Design Flow

The application of simulation tools is very important to improve the efficiency in system and circuit design. Various simulation tools exist on the market to support the design process. This chapter discusses topics that must be taken into account when selecting appropriate simulation tools.

As described in Section 2.2 the top-down design flow starts with the system concept which covers the complete system. The system is then divided into subcomponents down to the circuit and layout level. The choice of simulation tool depends on the design level addressed and the type of design (analog, RF, digital or mixed-signal). Simulators may cover more than one design level (Figure 3-1).

We distinguish between four categories of simulators, which are described in the following sections.

System level simulators

System level simulators provide efficient simulation algorithms to achieve a high simulation speed. This allows simulation of complete transmission systems containing a transmitter, channel and receiver with analog and digital parts. The simulation accuracy is restricted particularly for analog system parts. However, it allows the verification of system concepts. System modeling is supported by large libraries, which contain models of various system components, for example coders, modulators, and channels. The primary application of these tools is the system level design, also called concept engineering. They may also be partially used in block level design, for example to provide testbenches.

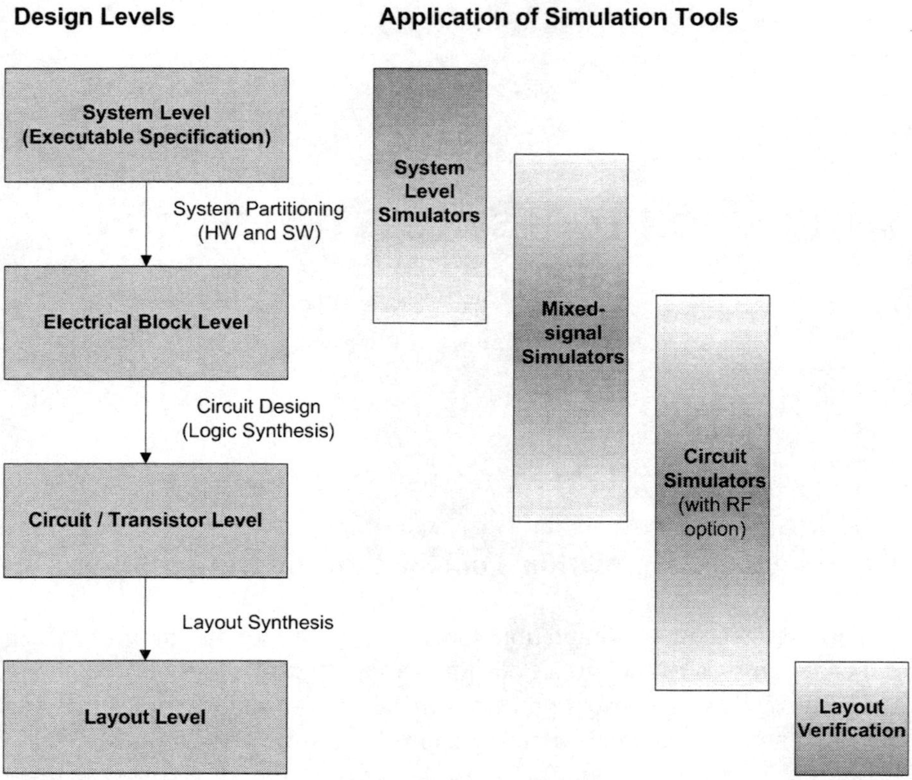

Figure 3-1. Simulation tool coverage in the mixed-signal design flow

Mixed-signal simulators

The main application of mixed-signal simulators is within the block level design where the partitioning into analog and digital hardware or software is performed. Mixed-signal simulation allows the common verification of analog and digital system parts, as well as the interfaces between them. Behavioral models are widely used at this design stage. The most important mixed-signal modeling languages are VHDL-AMS and Verilog-AMS.

The application of mixed-signal simulators can be extended to the system level if models of the system components exist. However, at present the model libraries of mixed-signal simulators do not achieve the complexity of the system level simulator libraries.

Mixed-signal simulators may also be used in circuit level design. In contrast to specialized RF circuit simulators they do not provide RF specific analyses.

Circuit level simulators

Most circuit level simulators support the simulation of circuit level descriptions (SPICE netlists) as well as analog behavioral models. Some simulators provide specialized simulation algorithms for the analysis of RF components (circuit envelope, periodic steady state for example). They provide an accurate analysis of components, but the simulation performance is too low to simulate large system parts.

With the ability to use behavioral models, circuit level simulators may also be used in block level design of analog subsystems. In addition layout effects can be included in circuit simulation by extraction of parasitics.

Layout verification

Layout verification is used to check if the design rules for a desired silicon technology are fulfilled. Layout effects (for example parasitic capacitances, substrate coupling) may be extracted and back annotated for circuit level simulation. The impact of layout and packaging on the desired circuit functionality can be analyzed. Layout verification is not discussed further.

Table 3-1. Overview of simulation tools

Simulator	Main design level	Additionally supported levels	Target	Examples
system simulator	system level	block level	complete system	ADS, CoCentric, MATLAB, SPW
mixed-signal simulator	block level	system level, circuit level	subsystems	ADVance MS, SMASH, AMS Designer, Saber
circuit simulator	circuit level	block level, (layout level)	blocks	Eldo, Spectre, Spice, ADS
layout simulator	layout level		components, packages	Assura, Calibre, Hercules

Some simulators and their application are outlined in Table 3-1. In some cases a co-simulation of different tools is used to accelerate the simulation, reuse models, or increase simulation accuracy. This topic is outlined in Chapter 12.

3.2 Specific Simulation Algorithms of RF Simulators

The traditional SPICE analyses are essential in analog circuit design. Their application to RF circuits may cause some problems resulting from the behavior of RF systems such as:

- The signals which are transmitted are narrowband signals. This means that a data signal with a relatively low bandwidth is transmitted at a very high carrier frequency. To simulate a sufficient portion of the data signal a large number of carrier waves must be simulated. This may exceed the performance of traditional transient analyses (memory and time consumption).
- RF receivers usually receive weak desired signals while large interference signals are present. This implies that the linearity of the receiver is a very important task for the designer requiring a precise simulation of nonlinearity.
- Improved transistor models are required to represent the behavior of RF transistors.

Specialized RF simulation algorithms are provided to improve the analysis of RF circuits and systems. They are available in RF simulators like ADS and SpectreRF but typically not in VHDL-AMS simulators. An exception is ADMS RF which combines ADVance MS and Eldo RF. The most important simulation algorithms are:

- Periodic Steady State Analysis (PSS)
- Harmonic Balance (HB)
- Transient Envelope Analyses (Envelope)

They provide a good accuracy for RF specific measurements at a sufficient simulation performance. The principle of these analyses is outlined in the following section.

Analysis for dynamic systems with weak nonlinearities

Different simulation algorithms can be used to analyze the frequency response of dynamic and nonlinear systems such as mixers and LNA's. The algorithms are:

- Periodic Steady State (PSS) in Cadence's SpectreRF Simulator
- Harmonic Balance (HB) in Agilent's ADS

The results of these analyses are the frequency spectra of the signals within the system including the wanted and unwanted harmonics (arising from nonlinearity).

The analysis is used to compute the steady state response of a nonlinear circuit, which is the response after the start-up transient has died down. The stimulus of the circuit is a limited number of sinusoidal signals. In the steady state, the system response is periodic according to the period length of the fundamental frequency. All input frequencies of the system must be an integer multiple of the fundamental frequency. The methods of computing the steady state solution are different in PSS and HB.

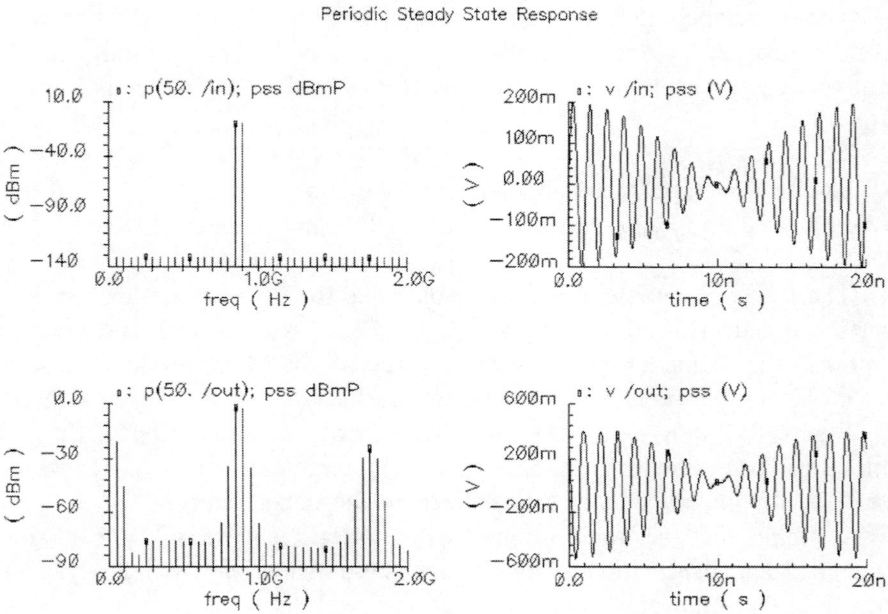

Figure 3-2. Results of a PSS analysis of an LNA

Figure 3-2 shows the results of a PSS analysis in frequency (left hand graphs) and time domain (right hand graphs). The input signal was two-tone with 850 and 900 MHz, each with a -10 dBm magnitude (upper graphs). Each input frequency must be an integer multiple of the fundamental frequency. Thus a fundamental frequency of 50 MHz is used in the example. This is equivalent to a period of 20 ns. To visualize frequencies up to 2 GHz, 40 harmonics of the fundamental frequency were computed. The time domain output of the LNA (bottom right hand graph) shows that the LNA is operated in the nonlinear area. The 3rd order harmonics at 800 MHz and 950 MHz are visible in the frequency domain (upper left hand graph). Other analyses are based on the steady state operating point, for example:

- periodic AC analysis
- periodic noise analysis
- periodic XF (periodic transfer function)
- periodic SP (periodic S-parameters)

The PSS analyses and the subsequent analyses are very important to determine the characteristics of RF systems and building blocks.

Transient envelope analyses

The envelope analyses address the narrow-band problem of wireless communication systems: signals with a relatively small bandwidth are transmitted at very high carrier frequencies. Transient envelope analyses are known as:

- Circuit Envelope Analysis (ADS from Agilent)
- Envelope Following Analysis (SpectreRF from Cadence)

The transient envelope analysis computes the envelope of a modulated carrier signal. This is demonstrated with a sine wave of 1 MHz, which is amplitude modulated on a carrier frequency of 900 MHz (modulation index 0.5). The simulation interval is 2 μs (two periods of the modulation signal).

Figure 3-3 shows the AM modulated carrier resulting from a transient analysis. To represent the modulated signal a large number of carrier periods must be computed, which is visualized in the detail interval (1…1.02 μs). This implies that the transient analysis is not efficient enough to evaluate a sufficient part of the modulation signal. The transient envelope analysis can speed-up the simulation of the modulation signal.

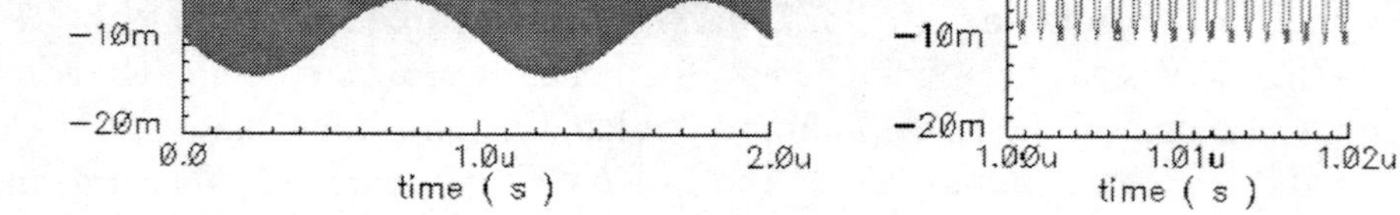

Figure 3-3. Results of traditional transient analyses (complete wave and detail)

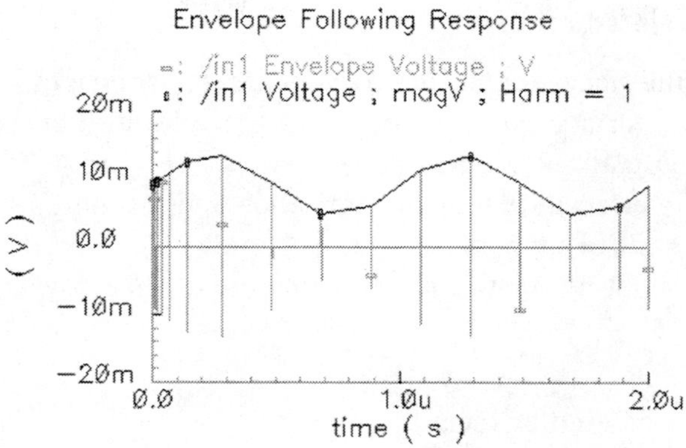

Figure 3-4. Result of the envelope following analysis (SpectreRF)

The envelope analysis was six times faster than the transient analysis of a small example LNA. The lower portion of the graph in Figure 3-4 shows the time domain signal of the modulated carrier. It can be seen, that the carrier signal is only partially computed. The black curve shows the envelope of the carrier which represents the modulating signal. There are too few sampling points to achieve a clear sine wave. The envelope analyses may be hardly applicable for multi-carrier or wideband modulation techniques.

3.3 Criteria of the Simulator Selection

A great number of simulation tools are on the market. This section presents some criteria which must be taken into consideration to identify the best simulation tool for a design task. The decision depends on the application, design flow, user interface, costs, and support.

Application related criteria

- In which design level(s) should the simulator be used?
- Which designs shall be mostly simulated (analog, mixed-signal)?
- Are special analyses needed (for example for RF)?
- Which model libraries are provided to speed-up the modeling of systems and testbenches?
- Is it possible to reuse models of former designs?
- Which simulation speed can be obtained?
- Is the size of the designs limited?

Design flow related criteria

- Are there interfaces for standardized modeling languages?
- Are there interfaces to other tools in the existing design flow (model import/export, simulator coupling)?
- Are there interfaces for tool customization and scripting?
- Is version control supported?
- Which computing platforms are supported (Windows, Unix, Linux, others)?

User interface related criteria

- Is a graphical user interface available?
- Schematic or netlist entry or both?
- Quality of documentation? (User guides, examples, reference manuals, tutorials, ...)

Cost related criteria

- Costs of licenses? (buying, leasing, public domain)
- Costs of support and version update?
- Time that is needed for user training?
- Costs of user training?
- Time/costs for software installation and maintenance?

Support Related Criteria

- Software support available?
- Web based support databases?
- Design service (special support on user applications)?

The criteria mentioned above shows that the selection of a simulation tool is very difficult. The integration of a new simulation tool often depends on the existing design flow. Some major vendors of EDA tools provide design frameworks where different tools have been integrated with a common user interface.

In the future, interfaces for standardized modeling languages, like VHDL-AMS, will simplify the exchange of models between simulators.

3.4 Internet Resources for Simulation Tools

The simulation tools mentioned in this chapter are continuously being improved. Latest information on supported features can be found on the internet. The list below shows the current tool vendors and the related internet addresses. The tools are assigned to the categories: system simulators, mixed-signal simulators, and analog RF simulators.

System Level Simulators

- Advanced Design System (ADS)
  Provider: Agilent Technologies
  http://eesof.tm.agilent.com/products/
- CoCentric System Studio
  Provider: Synopsys, Inc.
  http://www.synopsys.com/products/cocentric_studio/
- MATLAB
  Provider: The MathWorks, Inc.
  http://www.mathworks.com/products/matlab/
- Signal Processing Worksystem (SPW)
  Provider: CoWare
  http://www.coware.com
- APLAC
  Provider: APLAC Solutions
  http://www.aplac.com/

Mixed Signal Simulators

- ADVance MS
  Provider: Mentor Graphics
  http://www.mentor.com/ams/adms.html
- AMS Designer
  Provider: Cadence Design Systems
  http://www.cadence.com/products/
- SMASH
  Provider: Dolphin Integration
  http://www.dolphin.fr/medal/smash/smash_overview.html
- Saber
  Provider: Synopsys, Inc.
  http://www.synopsys.com/products/mixedsignal/saber/

Analog RF Simulators

- Advanced Design System (ADS)
  Provider: Agilent Technologies
  http://eesof.tm.agilent.com/products/
- Eldo RF
  Provider: Mentor Graphics
  http://www.mentor.com/ams/eldorf.html
- SpectreRF
  Provider: Cadence Design Systems
  http://www.cadence.com/products/

Chapter 4

SYSTEM LEVEL MODELING

4.1 System Level Simulation

The functionality of telecommunication systems has increased dramatically during recent years. The systems may support multiple standards and high data rates. Due to the cost reduction in chip production, modern digital transmission techniques are used. Sophisticated DSP routines (for example for protocols, error control coding, and modulation) provide high transmission quality in mobile systems.

Figure 4-1 shows the physical layer signal processing of a wireless local area network (WLAN) transmitter. The PDU train (protocol data unit) is a data stream, generated by the DLC (data link control) layer of HIPERLAN (High Performance Radio Local Area Network). Before the data is transmitted over a radio channel, algorithms including scrambling, FEC coding, and modulation are performed. In the receiver the reverse operations are used with additional algorithms for synchronization and channel equalization.

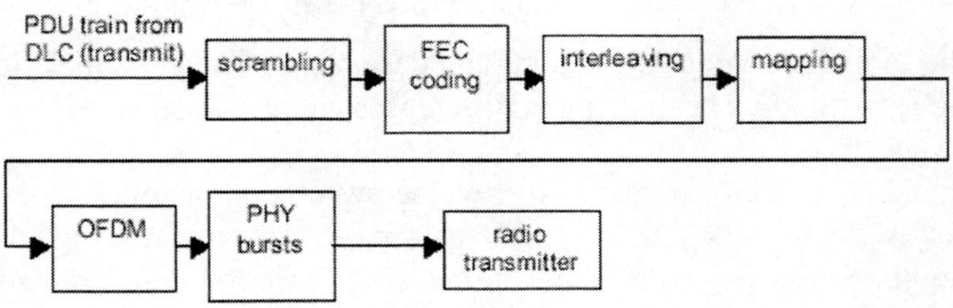

Figure 4-1. Physical layer of HIPERLAN/2 (transmitter)

System level simulation allows the evaluation of signal processing algorithms in the system environment. With validated reference libraries, the standard compatibility of algorithms can be evaluated as well as the overall system bit error rate (BER) over channel signal to noise ratio (SNR). The verified system level models are often used as reference for the implementation of algorithms.

Since system level simulators are designed to analyze large DSP systems, analog modeling is barely supported. On the other hand, it is important (with respect to System-on-Chip implementations) to investigate the impact of the RF subsystems on transmission system performance. The modeling of analog and RF components in system simulation is discussed in this chapter.

4.2 Simulation Technology of System Level Simulators

High level of abstraction

The simulation of whole transmission paths requires very fast simulation techniques. Therefore the models are often idealized:

- Models of DSP components represent the algorithm, but timing behavior is usually neglected.
- Analog system parts are sometimes completely neglected or they are modeled as ideal devices (for example an amplifier is often represented by a multiplication of a signal with a constant value).

For more accurate simulation of DSP components a co-simulation with a VHDL simulator has been provided for some years. This topic is not discussed here.

Due to higher transmission frequencies and more complicated radio transmission techniques the nonlinear behavior of the analog system part becomes more and more important for the system performance. Analog blocks can be modeled in spite of restrictions of system level simulators.

Distinctions between system level and mixed-signal simulators

The tools for analog, RF and system design use different simulation and modeling methods. The three main differences are discussed below.

1. Signals are often sampled: most of the system simulators (for example CoCentric or SPW) use equidistant samples to represent signals. The sampling rate for each signal is constant during simulation. Different sampling rates may be used for different signals or system parts. The user has to ensure that the sampling frequency is high enough to represent the

signal frequency without aliasing. Digital filter models H(z) must be used instead of analog ones H(s). This can increase the modeling error. Few tools (for example MATLAB and Ptolemy) provide time continuous data flow simulation.

2. Signals instead of nodes: system level simulators use signals, which cannot represent voltage and current as a conservative electrical node. Therefore it is difficult to model impedance mismatch between connected blocks. Often an ideal matching is assumed in system level simulation. More realistic port behavior can be achieved with additional modeling effort and parameters for port impedance.

3. No feedback between models: system level tools use a signal or data flow based simulation algorithm in a specified direction. There are only output and input ports; no bi-directional ports exist. A feedback between blocks must be modeled with additional ports and signals. The feedback loop must have a delay of at least one sample to enable correct simulation scheduling. In contrast an analog simulator solves the complete system at each step by iteration.

4.3 Complex Baseband Simulation

The very high value of the carrier frequency in wireless communication systems is the major problem in system simulation. It implies a very high sampling rate in simulation. The consequence is a low simulation performance, which results from a large number of iterations. Complex baseband modeling provides a more efficient simulation of RF subsystems.

4.3.1 Principle

Digital modulation techniques use magnitude r and phase ϕ of a carrier signal to transmit information. This means that the information does not depend on the carrier frequency value. The idea of baseband simulation is to transform the carrier frequency to zero. The advantage is that the required sampling rate now depends on signal bandwidth, not on carrier frequency (Figure 4-2).

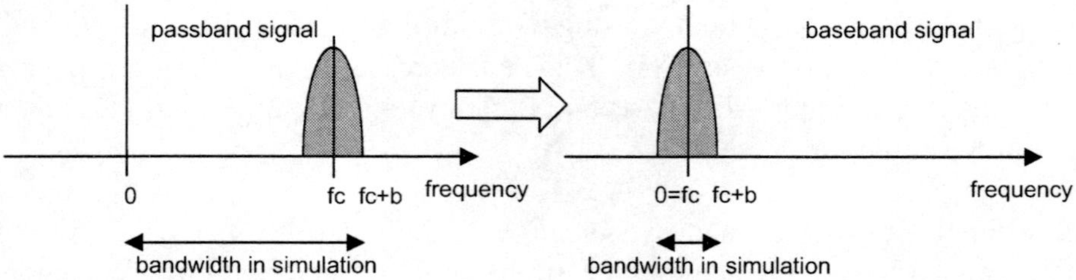

Figure 4-2. Passband and baseband representation of signals

creation of the quadrature representation

$$x(t) = a(t)\cos(\omega_0 t + \varphi(t))$$

$$x(t) = \underbrace{a(t)\cos(\varphi(t))}_{I(t)}\cos(\omega_0 t) - \underbrace{a(t)\sin(\varphi(t))}_{Q(t)}\sin(\omega_0 t)$$

$$x(t) = I(t)\cos(\omega_0 t) - Q(t)\sin(\omega_0 t)$$

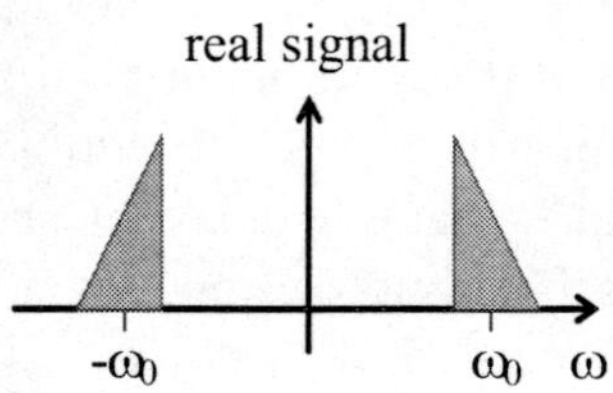

addition with the Hilbert transformed signal

$$\tilde{x}(t) = HT\{x(t)\}$$

$$\tilde{x}(t) = I(t)\sin(\omega_0 t) + Q(t)\cos(\omega_0 t)$$

$$y(t) = x(t) + j\tilde{x}(t)$$

$$y(t) = \{I(t) + jQ(t)\}\cdot e^{j\omega_0 t}$$

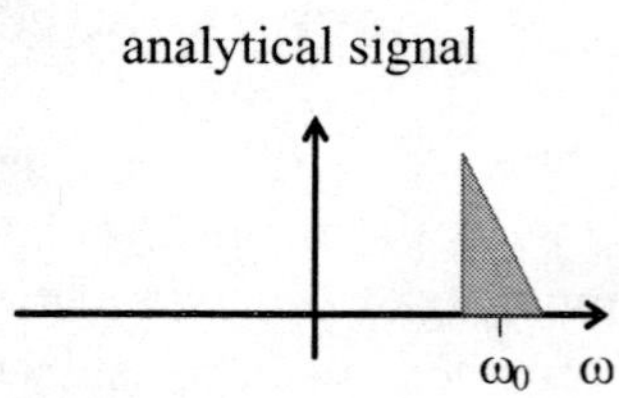

down-conversion into the complex baseband

$$z(t) = \{I(t) + jQ(t)\}\cdot e^{j\omega_0 t}\cdot e^{-j\omega_0 t}$$

$$z(t) = I(t) + jQ(t)$$

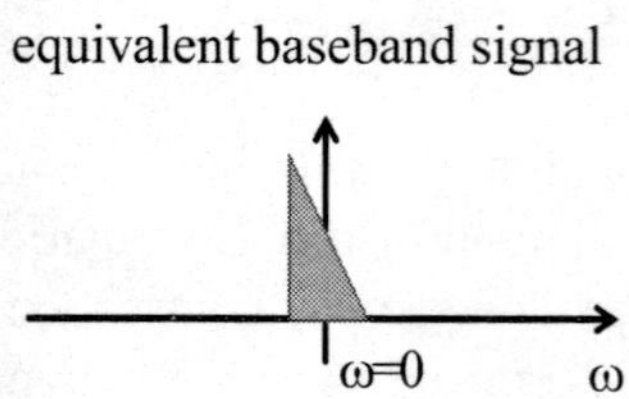

Figure 4-3. Signal transformation into the complex baseband

Figure 4-3 shows how the transformation of a modulated high-frequency carrier signal into the complex baseband can be carried out. The first part depicts the creation of the quadrature representation. The modulated carrier signal is a real signal, which contains positive and negative spectral components. The down-conversion in the complex baseband requires an analytical signal that contains no negative spectral components. For that purpose the Hilbert transformed signal of the real signal is built and added as an imaginary part to the real signal. The Hilbert transformation can simply be seen as a 90° phase shifter. In the last part of Figure 4-3 the analytical signal is down-converted into the complex baseband.

The equivalent baseband signal contains the amplitude- and phase-modulation information. It consists of two real signals, the inphase component I(t) and the quadrature component Q(t). The transfer functions of the RF blocks must also be transformed into the complex baseband. Because of the complex-valued signal the baseband models possess double the number of signal pins.

The baseband models influence the baseband signal (required signal) in their amplitude and their phase. Consequentially the following characteristics can be derived:

- AM/AM – amplitude to amplitude conversion
- AM/PM – amplitude to phase conversion
- PM/AM – phase to amplitude conversion
- PM/PM – phase to phase conversion

AM/AM and AM/PM conversion appears in all nonlinear, active RF components. The gain, the compression point, and the area of saturation can be read from the AM/AM curve. The AM/PM curve depicts the phase rotation, especially at strong input levels. The precise and efficient modeling of these characteristics is an important precondition for the system simulation of complex RF transmission systems. PM/AM and PM/PM conversions appear in modulators/demodulators and in certain mixing products. Additionally all mentioned characteristics can depend on frequency.

Noise is another important property to implement in baseband models. All noise characteristics have to be considered such as white noise, flicker noise, and phase noise. The superposition of different noise sources, filtered noise (colored noise) and large-signal modeling using random generators make efficient and precise noise modeling very difficult. Phase noise appears especially in autonomous blocks like oscillators, colored noise appears in amplifiers and mixers. Additionally, passband mixers shift the frequency of the noise. This frequency conversion is neglected in the baseband simulation.

4.3.2 Example for baseband simulation

The advantage of baseband modeling is illustrated in the wireless LAN system HIPERLAN/2 that transmits at a carrier frequency of approximately 5 GHz. It operates at two bands; the lower band from 5.150 GHz to 5.350 GHz, and the upper band from 5.470 GHz to 5.725 GHz. The bandwidth of the OFDM modulated signal is 20 MHz, split into 52 sub-carriers. Depending on the mode of operation, data rates from 6 Mbit/s to 36 Mbit/s are supported.

For safe data transmission, a raw bit error rate (BER) better than 1.0e-3 is required. To evaluate this, the transmission of approximately 10,000 bits is simulated. This complies with a transient analysis of 278 μs in 36 Mbit/s mode. Table 4-1 displays the simulation steps executed in passband and baseband simulation. In this example, the complex baseband simulation reduces the number of simulation steps by a factor of 250.

Table 4-1. Passband versus baseband simulation

	Passband simulation	Baseband simulation
highest signal frequency	carrier of about 5 GHz, sampled at 20 GHz	baseband bandwidth 20 MHz, sampled at 80 MHz
simulation step size	1.0/20 GHz = 50 ps	1.0/80 MHz = 12.5 ns
number of simulation steps	5.56 × 10e6	22.24 × 10e3

4.3.3 Restrictions and advantages of baseband modeling

In contrast to simulation with passband behavioral models, baseband simulation represents only spectral lines within a specified bandwidth around the carrier signal. Signal parts originating from nonlinear behavior outside this bandwidth are lost, for example harmonics of the carrier frequency. Unfortunately such effects could have an impact on the performance of subsequent receiver components. This is the main disadvantage of baseband modeling.

To improve simulation accuracy, an extended approach for baseband simulation is published in [Van00]. The multi-rate multi-carrier (MRMC) representation of signals uses a number of baseband signals at different frequencies and different bandwidths to represent a carrier signal. Harmonics of the carrier frequency can be considered in this way. This solution is not available as a commercial tool.

Because of the complex valued baseband signals, a baseband behavioral model cannot be replaced by a circuit level description of this block. Signal adapters, which convert from baseband to passband and vice-versa, are required to validate a circuit level model within a baseband test-bench.

Due to these restrictions baseband modeling must be used carefully. In full system simulation the speedup provided by this technology is crucial. It enables analysis of the impact of RF behavior on digital signal transmission.

4.4 Model Libraries for System Simulation

A feature of system level simulators is the availability of numerous models. They are used during concept engineering to simplify the development of system level models and test-benches. The system level simulators CoCentric System Studio and SPW specialize in the development of telecommunication applications. They provide large libraries with system components such as codec, error correction algorithms, modulators, filters, and more. Reference libraries are also available with models that are compatible with several communication standards. Table 4-2 shows a selection of models provided for the wireless communication domain.

Table 4-2. Sample reference libraries for wireless communication

CoCentric	SPW
Bluetooth	Bluetooth
GSM/GPRS	GSM and EDGE
cdma2000	WCDMA
DECT	Wireless LAN
IS-136	IS-136

The traditional application of system level simulators is development and verification of digital signal processing (DSP) algorithms. Therefore most of the library models belong to the DSP area. Nevertheless it is becoming more important to consider the imperfections of analog components in system level verification. In wireless systems, the analog components are concentrated in the RF front-ends of transmitters and receivers. Hence CoCentric and SPW provide a library to model RF front-ends.

The CoCentric RF library

The content of the CoCentric RF library is shown in Table 4-2.

Table 4-3. CoCentric RF library

Model	Description
ADConverter	analog to digital converter, nonlinear distortion
FrequencyDiv	divides the frequency of the input signal
FrequencyGen_QC	generates a frequency signal (complex)
FrequencySynt	frequency synthesizer
IQ_Mismatch	generates IQ amplitude and phase mismatch
Mixer_QC	RF mixer (complex)

Model	Description
NonLinAmp_QC	nonlinear RF amplifier (complex)
NonLinAmpS_QC	nonlinear RF amplifier, variable gain (complex)
Oscillator	oscillator with frequency/phase error, phase noise
PhaseComp	ideal comparison of the phases of two input signals
Vrms2dBm	converts root mean square voltage to dBm

The CoCentric RF library is designed for complex baseband modeling of the RF subsystems. The models include effects such as noise figure and intercept points. The models are coded in "C". The source code is not available.

The SPW RF library

In contrast to CoCentric, SPW provides models for complex baseband and passband signal representation. Table 4-4 shows the model groups contained in the SPW RF model library.

Table 4-4. SPW RF model library

Model category	Model
Amplifier	real, cascaded real (subtype of real amplifier) complex
Mixer	real, real lookup table based real with noise, complex
RF coupler	real, complex
Switches	ideal real switch (named select real) nonlinear real switch (named switch real) ideal complex switch (named select complex) nonlinear complex switch (named switch complex)
A/D converter	conditioning, midtread, simple midtread
Miscellaneous	dB gain real, dB gain complex (ideal amplification) pad real, pad complex (attenuation and noise) signal sign (returns sign of input signal) phase shift zero cross (detect zero crossing of input signal) ideal frequency multiplier (for complex signals) instantaneous frequency (simple frequency estimator) triggered sawtooth generator

These models allow the verification of transmitter and receiver front-end architectures with complex baseband or passband simulation techniques. Additionally Cadence provides J&K-models, which can represent complete receiver and amplifier RF-subsystems. J&K-models are black-box models configured with datasets generated automatically in an analog SpectreRF simulation.

4.5 Creation of Own Primitive and Hierarchical Models

System simulators provide a wide range of models, however it may be necessary to develop own models for user specific components or requirements.

There are generally two methods to create new models; both supported by CoCentric and SPW:

- hierarchical models
- primitive (custom coded) models

The development of hierarchical models requires no specific experiences. The user combines existing models in the schematic entry into a new component. The input and output signals are connected to port blocks. A symbol of the new component can be automatically created. Model parameters can be exported to the symbol if editing of the parameters on the top level is required.

If a block cannot be realized as a hierarchical model of existing blocks, many simulators provide interfaces to create custom coded blocks. Supported programming languages are "C", "C++" and "SystemC". The development of primitive models is more complicated than for hierarchical models. Some knowledge about the simulator's specific model interface (parameter and port signal access) is necessary. The simulator environment provides some support by generating model templates. An example is shown in the next section.

4.5.1 SPW modeling example

A low noise amplifier model is used to demonstrate how custom coded blocks are created in SPW. The development of primitive and hierarchical models is represented step by step in the SPW User Manual. The model shall have the parameters described in Table 4-5.

Table 4-5. LNA parameters

Parameter	Parameter name
Input resistance	Rin
Output resistance	Rout
Power gain	Gp
Noise figure	Fnoise
1 dB compression point	CP
3 dB corner frequency	fc
Sampling rate	fs

SPW system level model concept

Since SPW provides models for noise generation and frequency response, a hierarchical model is used. For the nonlinearity of the amplifier a primitive model will be created. The structure of the SPW model is depicted in Figure 4-4.

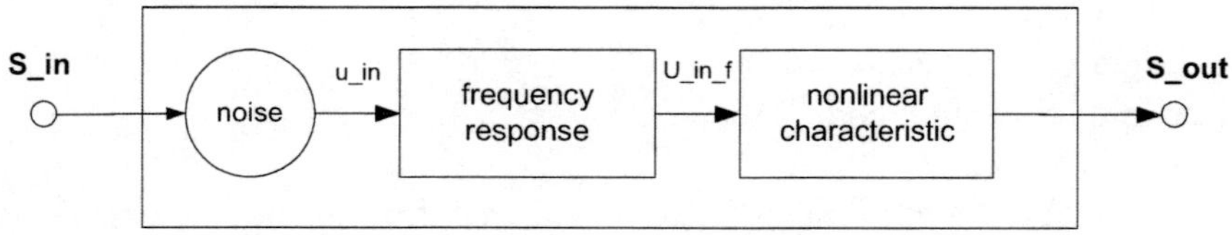

Figure 4-4. Structure of SPW LNA model

The primitive model for the nonlinear characteristic of LNA

Since an own primitive model for the nonlinearity of the LNA should be used, it must be created first. The model is named nonlin_amp.

The first step to create a primitive model is the creation of the symbol. An existing symbol can be opened in the schematic entry. After modification the symbol is saved with the new name in the user model library (Figure 4-5).

The model symbol describes the ports and the name of the model. The model parameters are specified in a parameter view, which is also created in the schematic entry.

Figure 4-6 depicts the parameter view of the model. The main parameters are used to configure the model functionality. Frequency response and noise figure are modeled hierarchically, therefore the parameters are not passed to this model. The miscellaneous parameter section contains information for the error and overflow handling of the simulator. It must not be changed.

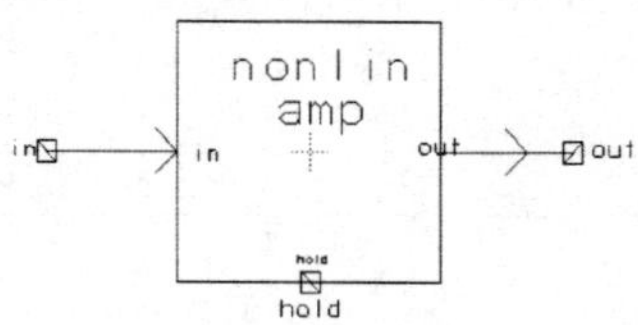

Figure 4-5. Symbol of the primitive model nonlin_amp

COMPLEX NONLINEAR AMPLIFIER BLOCK PARAMETERS	
MAIN PARAMETERS:	
Sampling frequency	1.0e6
Gain in dB	10.0
1 dB compression point (dBm)	15.0
Input resistance	50.0
Output resistance	50.0
MISCELLANEOUS PARAMETERS:	
Initial value	0.0
Overflow value	0.0
Error count before action	1
Action taken (stop or continue)	'stop'

Figure 4-6. Parameter view of nonlin_amp

After creation of the symbol and parameter views of the model, the source code is created. It is supported by the block wizard, which is launched from the Symbol Schematic window. The wizard creates templates for the C source code and the header file. They include interfaces for ports and parameters used in the model. (This is the reason why the symbol and the parameter view must be defined first.)

With the "View Header" and "View C Code" buttons, the files are opened in a text editor for editing. After saving the files, the model is compiled by clicking on the "Compile" button. The message area indicates success or errors of compilation. If necessary, "View Header" and "View C Code" are used again to change the model code. After successful compilation, the model is ready for simulation.

The following presents the method used to modify the templates. In the header file the interface of the model is completely defined. Since two state variables are used in the model, their definition must be added in the header file. (State variables retain their values between the model calls in simulation.)

```
STRUCT St_nonlin_amp_my_lib {
  int instance;
  double k1;
  double k2;
};

#define S_k1 (spb_state->k1)
#define S_k2 (spb_state->k2)
```

The lines marked bold are added to define the state variables. The macro #define ... is optional. It makes use of the state variables easier.

The source code file contains the model functionality. It contains three functions:

- initialize function
- run output function
- termination function

The function interface is completely defined, but the signal processing operations have to be added. The initialize function is executed at the start of each simulation. In this example it is used to compute the coefficients for the tanh nonlinearity from the model parameters. The code is shown below.

```
int In_nonlin_amp_my_lib(spb_parm, spb_input, spb_output,
spb_state)
STRUCT Pt_nonlin_amp_my_lib *spb_parm;
STRUCT It_nonlin_amp_my_lib *spb_input;
STRUCT Ot_nonlin_amp_my_lib *spb_output;
STRUCT St_nonlin_amp_my_lib *spb_state;
{
   double cp_lin,gp_lin,cp_temp;
   cp_temp =  P_comp_pt - (P_gain+3.0) + 10.0; /*input ref. CP*/
   cp_lin   = pow(10.0, (cp_temp-30.0)/10.0);
   gp_lin   = pow(10.0, (P_gain+3.0)/10.0);

   S_k1        = 1.0/0.504 / sqrt(cp_lin);
   S_k2        = 1.0/0.504 * sqrt(gp_lin * cp_lin);
  return (SYS_OK);
}
```

The bold text parts are added. The algorithm is adapted from an analog behavioral model and will not be discussed here. The states *k1* and *k2* (represented by S_k1 and S_k2) are used in the run output function.

```
int Ro_nonlin_amp_my_lib(spb_parm, spb_input, spb_output,
spb_state)
STRUCT Pt_nonlin_amp_my_lib *spb_parm;
STRUCT It_nonlin_amp_my_lib *spb_input;
STRUCT Ot_nonlin_amp_my_lib *spb_output;
STRUCT St_nonlin_amp_my_lib *spb_state;
{

  O_out = 1.0 * S_k2 * tanh(I_in * S_k1);

  return(SYS_OK);
}
```

The pre-computed coefficients make it easy to compute the model output *O_out* from the input signal *I_in*.

The termination function is not changed because the model does not require computations at the end of simulation.

The hierarchical model lna_new

The primitive model nonlin_amp is now used in the hierarchical LNA model. A *detail* view is created in the schematic entry. It consists of a noise generator and a lowpass filter from the SPW model libraries, together with our nonlinear amplifier. It is depicted in Figure 4-7. The parameters of the model are defined in the box. If the model is instantiated their values can be specified in the symbol. From that schematic a symbol view can be automatically created. After this the new model *lna_new* is complete and can be used in simulation.

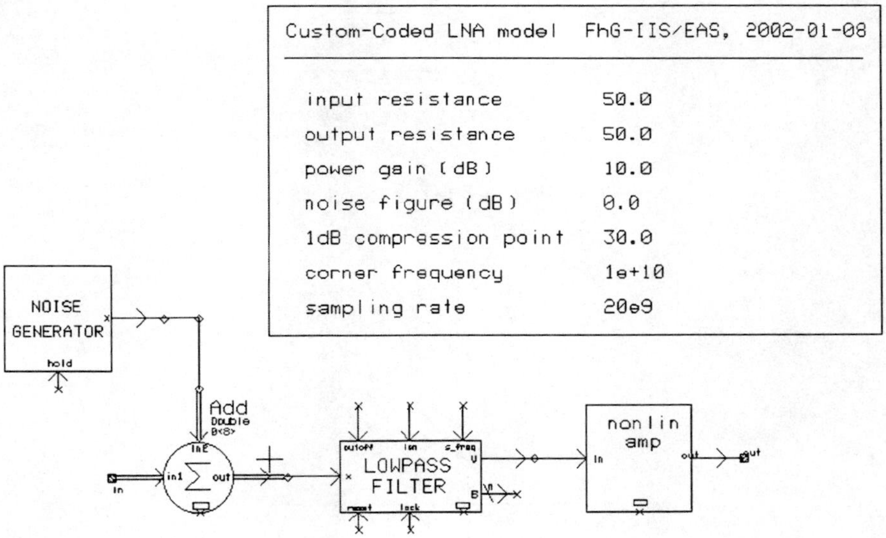

Figure 4-7. Hierarchical model lna_new

Model test-bench

The created model *lna_new* is inserted in a test-bench to verify gain and compression point of the model (Figure 4-8). It creates a one tone test signal with a power sweep. The input and output power of the LNA are measured. The simulation results are visualized with the SPW Signal Calculator (Figure 4-9).

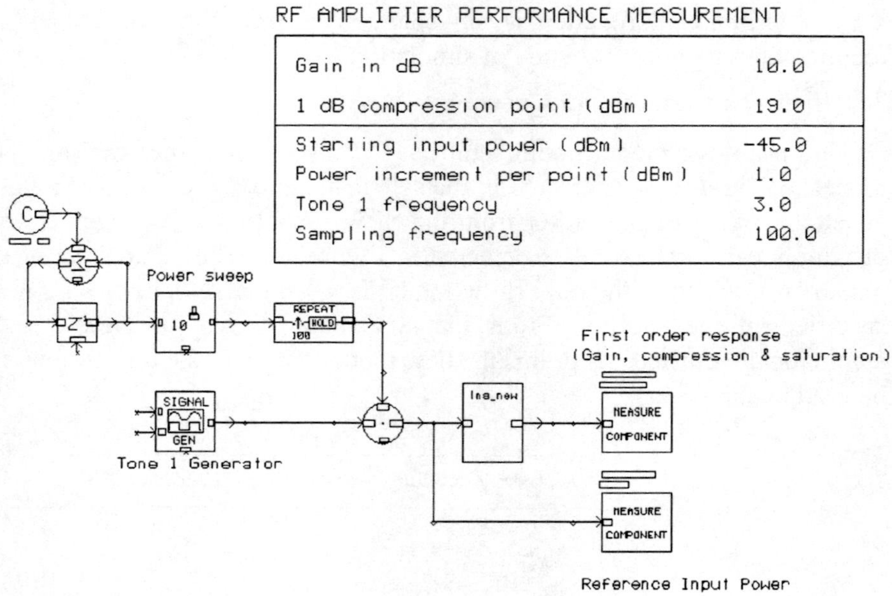

Figure 4-8. Test-bench for LNA compression point

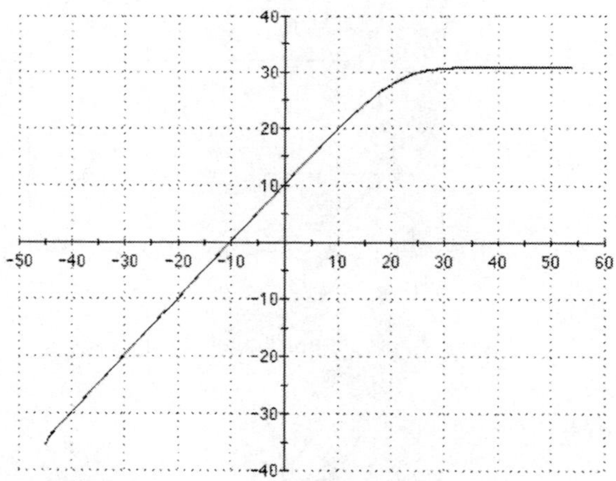

Figure 4-9. lna_new simulation result output power versus input power

Chapter 5

VHDL-AMS FOR BLOCK LEVEL SIMULATION

5.1 Introduction

Current telecommunication circuits consist of digital and analog blocks. In order to describe and simulate these circuits a behavioral description language is required that covers both levels and the interaction between them. A language that fulfills these expectations is VHDL-AMS.

VHDL-AMS is a hardware description language for the description and simulation of digital, analog, and mixed-signal systems. The language was standardized by the Institute of Electrical and Electronic Engineers (IEEE) as *IEEE Standard 1076.1-1999, VHDL Standard Analog and Mixed-Signal Extensions* [Std99]. It is a strict superset of the digital VHDL IEEE Std 1076-1993. The VHDL-AMS standard supports the development of tool-independent models. Currently, a number of VHDL-AMS simulation engines are available (see for example [MGC04]). VHDL-AMS can be used in different phases of the design and simulation flow. It is applicable in the top-down phase as well as during bottom-up verification (see also [ChB99], [APT04]).

In this and the following chapter we will give a brief introduction to VHDL-AMS and its usage in telecommunication applications. These explanations should help you to

- Gain an overview of what kind of modeling and simulation tasks can be solved using VHDL-AMS
- Understand and apply existing VHDL-AMS models
- Change and refine existing models
- Develop from scratch more complicated or less complicated models

The text is addressed to readers with a basic knowledge of digital and analog simulation. Some knowledge of the digital VHDL 1076-1993 language is helpful but not absolutely necessary to gain a first impression of the language. We will present selected models that can be applied to RF design together with special modeling methods and their application using some complex examples.

5.2 VHDL-AMS Standardization

The hardware description language VHDL was standardized in 1987. The language was originally developed to describe large digital integrated circuits (ICs) in a unified way. VHDL is an abbreviation for VHSIC Hardware Description Language. VHSIC is an abbreviation for Very High Speed Integrated Circuits. Users soon appreciated the advantages of the language for modeling, documentation, and simulation of simple or complex digital systems. As a result of the language's standardization the exchange of VHDL models, developed by different users and for different tools, was facilitated. The standard accelerated the development of simulation, and also synthesis, tools based on VHDL. This is a benefit from the users' point of view as well as from the EDA companies' position. Besides this, VHDL allows

- Structural descriptions of digital systems
- Behavioral descriptions of basic building blocks

That means users can define their own primitives for the structural descriptions in an easy way. This gives them great flexibility to describe their system ideas. These possibilities are of particular value if a system specification has to be simulated.

All these advantages were not applicable for analog designers until the beginning of the 1990's. Very powerful network simulation engines based on SPICE simulation ideas were available [Kun95]. However, a standard for behavioral models and a link to digital simulation engines did not exist. This was the starting point for the definition of VHDL-AMS.

The IEEE 1076.1 Working Group tasked to overcome these limitations was created under the auspice of the IEEE Design Automation Standards Committee (DASC). The task was to develop analog and mixed-signal extensions to the VHDL language. As a result of the activities of the working group, the *1076.1-1999 IEEE Standard VHDL Analog and Mixed-Signal Extensions* was approved in 1999. This Language Reference Manual is available from the IEEE. The VHDL 1076.1 language is informally known as VHDL-AMS, where AMS is an abbreviation for Analog and Mixed-Signal.

VHDL-AMS is a strict extension of the digital VHDL 1076-1993 language. Thus, each VHDL model is also a VHDL-AMS model. New language constructs for the description of continuous behavior over time and frequency are smoothly included in VHDL.

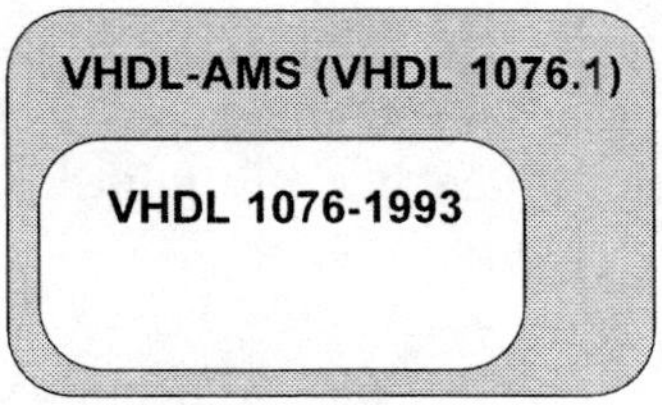

Figure 5-1. VHDL-AMS as a superset of VHDL 1076-1993

VHDL-AMS is a language to model and simulate digital, analog, and mixed-signal systems in a unified way. All the organizational capabilities of VHDL remain valid.

The IEEE 1076.1 Working Group that promoted the VHDL-AMS development reacts to experiences of the VHDL-AMS deployment. Last activities included, for instance, the development of standard packages for multiple domain (that is electrical/non-electrical) simulation. The balloting concerning *IEEE P1076.1.1 Standard VHDL Analog and Mixed-Signal Extensions – Packages for Multiple Energy Domain Support* [Std03] was carried out in 2004. More information is available online from the IEEE 1076.1 Working Group website:

http://www.vhdl.org/analog

5.3 A Simple Block Level Example – Analog PLL

One of the advantages of VHDL-AMS is the description of circuit blocks at a high level of abstraction. This helps to determine and check basic system parameters without wasting too much effort considering second order effects at the beginning of a design process. Thus, the idea is to use an executable specification at the beginning of a top-down design. The high level of abstraction helps to reduce simulation times. System parameters and structures can be easily modified.

As an example, we start with modeling a simple analog PLL (Phase Locked Loop) in the passband circuit using VHDL-AMS. The starting point to model the blocks is a description of their functionality described by mathematical expressions. Prior knowledge of the realization of the blocks at the transistor level is not required. Later on in the design process, models can be replaced by more detailed descriptions. The interface of the blocks will not be changed in this procedure. Only the description of their working mechanisms will become more complicated. That means we have to use pin-compatible models from the beginning.

5.3.1 Mathematical models of basic blocks

The behavior of a PLL can be illustrated using a simple description. Instead of circuit models of the basic building blocks we use behavioral descriptions of

- Signal source described by a series connection of a voltage source and a voltage-controlled oscillator
- Phase detector realized by a multiplier
- Ideal first order lowpass filter
- Voltage controlled oscillator

The output of the signal source is a frequency-modulated signal. Demodulation is carried out by the PLL [PeD91], [Kam92]. Figure 5-2 represents the schematic of the PLL.

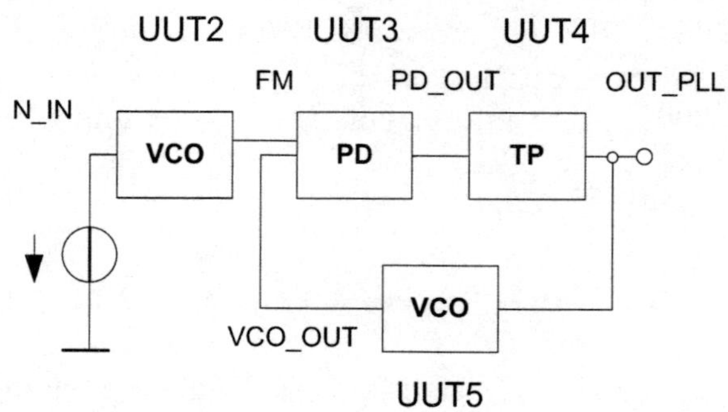

Figure 5-2. PLL circuit

The blocks may be realized by ideal voltage-controlled voltage sources. Parameters of the models are introduced in the following. The FM input signal is demodulated by the PLL. The output voltage should follow the input voltage.

Signal Source

An independent voltage source drives the voltage controlled oscillator UUT2. The output of the VCO is a frequency-modulated voltage. It is used as PLL input.

Phase detector (PD)

The phase detector is realized in a simple manner. The values of the two input waveforms are multiplied by the gain of the phase detector. Two open input branches, which are used to measure the input voltages vin_1 and vin_2, and an ideal controlled voltage source that drives the output can be used to model the phase detector:

$$vout(t) = gain \cdot vin_1(t) \cdot vin_2(t)$$

Voltage controlled oscillator (VCO)

The frequency of the sinusoidal VCO output voltage depends on the input voltage *vin* measured by an open branch. In our simple model we assume that the frequency is proportional to the input voltage plus center frequency f_0. Considering the basic definition that the derivative of the phase φ equals the product of frequency and 2π we start with

$$\frac{d\varphi}{dt} = 2\pi \cdot \left(kf \cdot vin(t) + f_0\right) \quad \text{and the initial condition } \varphi(0) = \varphi_0$$

The output voltage results from

$$vout(t) = Ampl \cdot \sin\left(\varphi(t)\right)$$

kf is the VCO gain measured in Hz/V. *Ampl* is the amplitude of the output voltage.

Lowpass filter (LP)

The lowpass filter is characterized by its cut-off frequency and gain. The input voltage *vin* is measured by an open branch. It controls the voltage *vout* of an ideal voltage source. The Laplace transfer function of the lowpass filter is given by

$$H(s) = \frac{gain}{1 + \frac{s}{2\pi \cdot f_c}} = \frac{vout(s)}{vin(s)}, \text{ i.e. } vout(s) = vin(s) \cdot \frac{gain}{1 + \frac{s}{2\pi \cdot f_c}}.$$

5.3.2 Structural description of the PLL circuit in VHDL-AMS

In order to simulate the PLL circuit these mathematical models have to be translated into VHDL-AMS descriptions. At the beginning, we assume that the models of the basic building blocks are available. We only have to connect and parameterize them. The structural description can be done with reference to the interface descriptions of the block models. The interfaces are described in VHDL by the entity declarations. These declarations contain the identifiers of the generic model parameters, their types, and optionally, their default values. The connection points are summarized in the port list. The declaration of a connection point characterizes an element of the port list. A terminal, for instance, is a connection point in a network model. Furthermore, identifiers and a characterization by type or nature belong to a port declaration. In the case of an electrical network the nature is ELECTRICAL.

The entity declaration of the voltage source model used is shown in the following lines of code. The parameter WAVE describes the voltage waveform as a list of times and values in a similar way as in any well-known SPICE simulator [QNP93].

```
entity VPWL is
   generic  (
      WAVE    : REAL_VECTOR;       -- time value pairs T1, V1, …
                                   -- units: [s] and [V]
      ACMAG   : REAL := 0.0;       -- AC magnitude
      ACPHASE : REAL := 0.0        -- AC phase
      );
   port     (
      terminal P : ELECTRICAL;     -- positive terminal
      terminal N : ELECTRICAL      -- negative terminal
      );
end entity VPWL;
```

The VCO entity is declared in a similar way. The assert statement checks whether model parameters are assigned in a correct way during instantiation.

```
entity VCO is
   generic (F0           : REAL := 1.0; -- center frequency [Hz]
            KF           : REAL := 1.0; -- gain [Hz/V]
            AMPL         : REAL := 1.0; -- amplitude [V]
            PHI0         : REAL := 0.0  -- initial phase [rad]
           );
   port    (terminal INP  : ELECTRICAL;  -- input terminal
            terminal OUTP : ELECTRICAL   -- output terminal
           );
begin
   assert F0 > 0.0 and KF > 0.0
     report "F0 and KF > 0.0 required."
     severity ERROR;
end entity VCO;
```

We assume that the other entity descriptions are available in the same way. A PLL netlist in accordance with Figure 5-2 can then be described in the following way.

```
library IEEE;
use IEEE.ELECTRICAL_SYSTEMS.all;
use IEEE.MATH_REAL.all;
use WORK.all;

entity BENCH is end entity BENCH;

architecture PLL of BENCH is
   terminal N_IN, FM, PD_OUT, VCO_OUT, OUT_PLL : ELECTRICAL;
begin

V1: entity VPWL(SPICE)
      generic map (WAVE =>
        (0.0, -1.0, 50.0E-6, -1.0, 150.0E-6, 1.0, 200.0E-6, 1.0))
      --(time1, value1, time2, value2, ...)
      port map (P => N_IN, N => ELECTRICAL_REF);

UUT2: entity VCO(BASIC)
      generic map (F0 => 1.0E6, KF => 100.0E3,
                   AMPL => 4.0, PHI0 => -MATH_PI/2.0)
      port map (INP => N_IN, OUTP => FM);

UUT3: entity PD(BASIC)
      generic map (GAIN => 2.5)
      port map (IN1 => FM, IN2 => VCO_OUT, OUTP => PD_OUT);

UUT4: entity FILTER(LP)
      generic map (FC => 20.0E3)
      port map (INP => PD_OUT, OUTP => OUT_PLL);

UUT5: entity VCO(BASIC)
      generic map (F0 => 1.0E6, KF => 100.0E3)
      port map (INP => OUT_PLL, OUTP => VCO_OUT);

end architecture PLL;
```

Thus, existing VHDL-AMS models can be used to describe new simulation tasks. The description starts with a context clause. The first two lines allow access to the nature ELECTRICAL. Our models were compiled into the logical library WORK. The third line is included in the model to enable instantiation of design entities that were compiled into the library WORK. An empty entity declaration follows as there are no connection points to higher hierarchy levels at the top.

The structural description follows in the architecture PLL. At the beginning, the internal nodes N_IN, FM, and so on are declared. After the reserved word **begin**, the instantiations of the models follow. There are different methods to instantiate models in VHDL-AMS. Here we use the direct instantiation method. The **generic map** associates actual values with model parameters. If the actual value and the default value of the entity declaration are equal, a value assignment is not necessary. Note that only parameter values, and no units, are used. In general, models should always be written in a way that the usage of SI units can be assumed. However, ultimately the user still must check whether this assumption is met. As a consequence, a clear documentation of units in the source code of models helps to avoid misunderstandings. The port map describes how to connect the ports to the nodes of the circuits. In this example we use a named association. That means the identifiers of the entity declarations are used in the mapping list. We note that each instantiation must start with a label (for example UUT2, UUT3,...) and is continued by the reserved word **entity** followed by the entity name and the architecture name enclosed within parentheses.

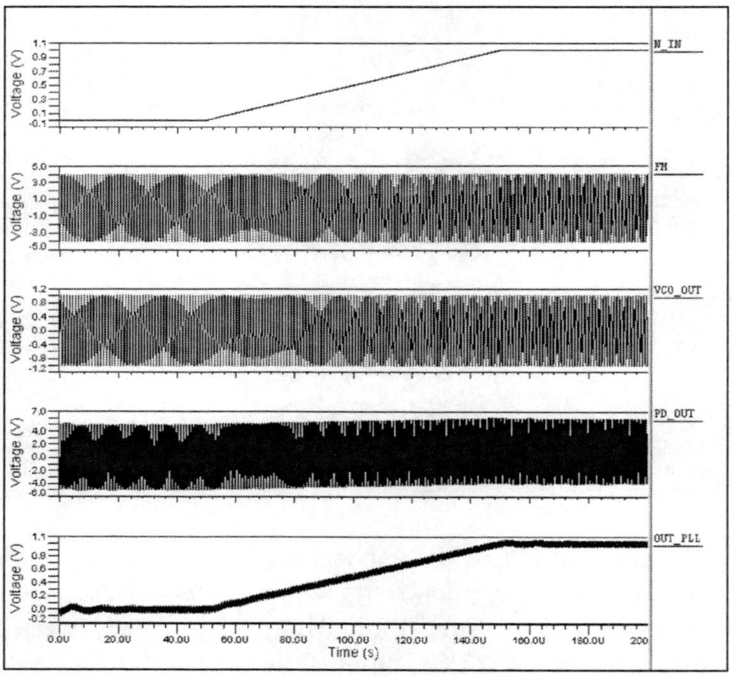

Figure 5-3. Results of a passband simulation to 200 μs

Different model descriptions can be associated with the same entity. This functionality is described in VHDL-AMS in architecture bodies. The name of the architecture used follows the entity name in the instantiation statement enclosed within parentheses. Last but not least we mention that the electrical reference node is named ELECTRICAL_REF in the nature made available by the context clause. Figure 5-3 shows the results of the PLL simulation.

The input voltage is measured at node N_IN. The result of the demodulation is available at node OUT_PLL.

5.3.3 VHDL-AMS description of basic blocks

In the previous section we explained how to use existing models. Let us now look at three of these models. One of the advantages of VHDL-AMS is that you can write your own models in a similar manner.

Phase detector

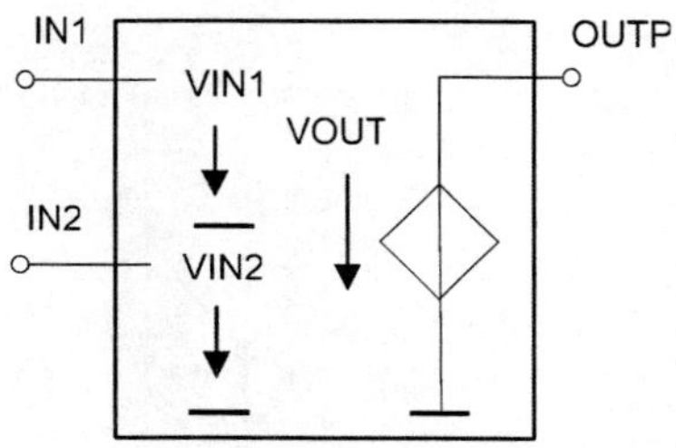

Figure 5-4. Structure of the basic phase detector model

The internal structure of the phase detector model is shown in Figure 5-4. The following code lines describe the VHDL-AMS model:

```
library IEEE;
use IEEE.ELECTRICAL_SYSTEMS.all;

entity PD is
   generic (GAIN          : REAL := 1.0    -- gain
           );
   port    (terminal IN1  : ELECTRICAL;    -- first input
            terminal IN2  : ELECTRICAL;    -- second input
            terminal OUTP : ELECTRICAL     -- output terminal
           );
end entity PD;

architecture BASIC of PD is
  quantity VIN1 across IN1;                -- 1 st open input branch
  quantity VIN2 across IN2;                -- 2 nd open input branch
  quantity VOUT across IOUT through OUTP; -- output branch
begin
  VOUT == GAIN*VIN1*VIN2;                  -- see PD description
end architecture BASIC;
```

The entity declaration describes the model interface. The quantity statements in the architecture BASIC declare branches. A probe branch that measures the voltage VIN1 between the terminal IN1 and the electrical reference is declared by `quantity VIN1 across IN1;` which is an open branch. The branch current is zero. The second input branch is declared in the same way. The third quantity statement presents the output voltage source branch. It connects the terminal OUTP and the electrical reference node. The branch voltage is VOUT and the branch current IOUT. The value of the branch current depends on the interconnection of the phase detector with other models. The relationship that has to be fulfilled by the branch voltages and currents is given by the simultaneous statement `VOUT == GAIN*VIN1*VIN2;` Thus, the VHDL-AMS model represents the mathematical model description given in Section 5.3.1.

Voltage controlled oscillator

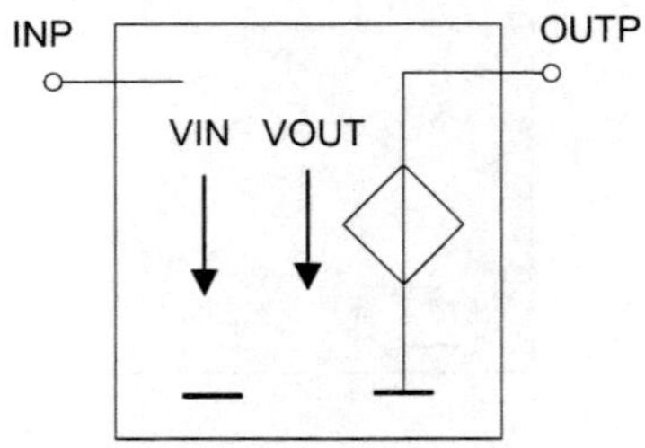

Figure 5-5. Structure of the basic VCO model

The VCO model is written in a similar way to the phase detector. The entity declaration was already shown in Section 5.3.2. The architecture contains an additional declaration for PHI that is neither a branch current nor a branch voltage. Furthermore the initial condition is considered during operating point analysis (DOMAIN equals QUIESCENT_DOMAIN). PHI'DOT is the time derivative of PHI. The SIN function and the mathematical constants can be found in the package MATH_REAL that is introduced with the first two lines of the following code:

```
library IEEE;
use IEEE.MATH_REAL.all;

architecture BASIC of VCO is
  quantity VIN  across INP;                 -- open input branch
  quantity VOUT across IOUT through OUTP; -- output branch
  quantity PHI  : REAL;                     -- free quantity PHI
begin
  if DOMAIN = QUIESCENT_DOMAIN use
     PHI == PHI0;                           -- initial condition
  else
     PHI'DOT == MATH_2_PI*(KF*VIN + F0);   -- see VCO description
  end use;
  VOUT == AMPL*SIN(PHI);                    -- see VCO description
end architecture BASIC;
```

Lowpass filter

VHDL-AMS provides the 'LTF attribute to describe the Laplace transfer function. This attribute can be applied to quantities, such as analog waveforms. Analog waveforms are, for example, branch voltages and currents. Thus, the main functionality of the lowpass filter is given by a simultaneous statement of the form `VOUT == VIN'LTF(numerator, denominator)`. Parameters of the 'LTF attribute are real arrays for the numerator and denominator of the Laplace transfer function that contain the coefficients of s^i of numerator and denominator respectively. The two arrays can be described using the following form: (0 => coefficient of s^0, 1 => coefficient of s^1,…). The internal structure and the interface of the model are similar to the VCO. Thus, we get the following model:

```
library IEEE;
use IEEE.ELECTRICAL_SYSTEMS.all;
use IEEE.MATH_REAL.all;

entity FILTER is
   generic (GAIN         : REAL := 1.0; -- gain
            FC           : REAL := 1.0; -- cut-off frequency [Hz]
            ZF           : REAL := 1.0  -- zero frequency [Hz]
           );
   port    (terminal INP  : ELECTRICAL;   -- input terminal
            terminal OUTP : ELECTRICAL);  -- output terminal
begin
   assert FC > 0.0 and ZF > 0.0           -- parameter conditions
     report "FC and ZF > 0.0 required."
     severity ERROR;
end entity FILTER;

architecture LP of FILTER is                -- lowpass description
  quantity VIN  across INP;                 -- open input branch
  quantity VOUT across IOUT through OUTP;   -- output branch
begin
  VOUT == VIN'LTF((0 => GAIN), (0 => 1.0, 1 => 1.0/MATH_2_PI/FC));
end architecture LP;
```

If a pole-zero filter should be used instead of the lowpass filter, a second architecture PZ can be combined with the FILTER entity.

```
architecture PZ of FILTER is
  quantity VIN  across INP;                  -- input branch
  quantity VOUT across IOUT through OUTP;    -- output branch
begin
  VOUT == GAIN*VIN'LTF(
         (0 => 1.0, 1 => 1.0/MATH_2_PI/ZF),  -- numerator
         (0 => 1.0, 1 => 1.0/MATH_2_PI/FC)   -- denominator
          );
end architecture PZ;
```

The architecture PZ realizes in the Laplace domain

$$vout(s) = gain \cdot vin(s) \cdot \frac{1 + \dfrac{s}{2\pi \cdot zf}}{1 + \dfrac{s}{2\pi \cdot fc}}.$$

This architecture PZ can be instantiated in the PLL description instead of the lowpass filter architecture LP.

5.4 Summary

This chapter provided a short overview of the features of the VHDL-AMS behavioral description language, which can be used for modeling of digital, analog, and mixed-signal systems. We demonstrated how the language could be used for block simulation at a high level of abstraction. The starting point is always a mathematical description of the behavior of the blocks. VHDL-AMS offers powerful statements to translate a mathematical model of a block into a model that can be evaluated by a simulation tool. Some of the basic language features were introduced and used in an informal manner in this chapter. Our aim was to provide an idea of how modeling with VHDL-AMS works. However, to exploit the whole power of the language a systematic approach to the language is essential. An introduction to VHDL-AMS follows in the next chapter.

Chapter 6

INTRODUCTION TO VHDL-AMS

6.1 Aim of this Introduction

In this chapter, we will introduce the fundamental concepts of VHDL-AMS. This will help to understand existing VHDL-AMS models and will open the opportunity to modify models and develop new ones. We cannot explain all details of the language. Instead, we will touch on the main ideas and practices to form a basic understanding useful for the beginning (see also [ChB99]). The chapter is subdivided into the following topics:

- Repetition of VHDL 1076-1993, which is the pure digital predecessor of VHDL-AMS
- Description of conservative systems where analog blocks of arbitrary nature can be described by means of ordinary equations and algebraic equations
- Description of nonconservative systems where analog blocks can be described by means of signal flow diagrams
- Mixed-signal simulation, including basic knowledge concerning the combination of analog and digital simulation models
- Analysis domains, i.e. additional frequency domain simulation modes in VHDL-AMS
- Further features of VHDL-AMS as an outlook on advanced modeling techniques.

In the following, the general syntax of VHDL-AMS statements is described using a simple variant of Backus Naur Form (BNF), similar to those used in the VHDL-AMS standard [Std99].

6.2 Repetition of Basics of VHDL 1076-1993

VHDL 1076-1993 is widely used in electronic design. Most of the designers are familiar with this language to describe and simulate their digital design.

VHDL-AMS is a strict superset of VHDL 1076-1993. As a preliminary, we will look at some basic ideas of digital VHDL and then explain the analog extension. We do not provide a detailed introduction into digital VHDL, but we will touch on the main topics. More information on VHDL 1076-1993 can be found in the books [Nav93], [Ash02] as well as online:

- The Hamburg VHDL Archive
  http://tech-www.informatik.uni-hamburg.de/vhdl

In the following, we will mainly address VHDL and will consider some extensions to VHDL-AMS where applicable.

6.2.1 Design units

The basic organizational unit in VHDL-AMS is the design unit. There are two classes of design units: primary and secondary. The *primary design units* are:

- **entity**
- **package**
- **configuration**

Secondary design units depend on primary design units to some extent. The secondary design units are

- **architecture**
- **package body**

VHDL-AMS allows building up a system model in a hierarchical way with the help of building blocks. The description of the interface of a building block together with its associated structural or behavioral description defines a design entity. A *design entity* represents a model of a building block. It is provided by a description of its interface and an associated architecture. It is possible to define different models for the same building block. All these models consist of the same interface but with different architectures. A first example was provided at the end of the last chapter.

Design unit entity

An **entity** declaration is the description of the interface between a given design entity and the environment in which it is used. In VHDL-AMS, the connection points of a building block are called ports. An entity declaration describes the ports and parameters of a building block.

The simplified form of an entity declaration is

```
entity entity_name is
     generic    (parameter_list);
     port    (port_list);
end entity entity_name;
```

The parameter list defines names and types of constants that can be used in an associated architecture. The generic constants can be initialized in the entity declaration. The default values can be overwritten during instantiation.

The port list defines names, directions and types of channels for communication between a building block and its environment. These are

- **signal** ports for digital waveforms with modes **in**, **out**, **inout**, or **buffer** (known from VHDL)
- **quantity** ports for analog waveforms with modes **in** or **out** (new in VHDL-AMS)
- **terminal** ports for conservative connection points that carry analog flow and across waveforms without direction (new in VHDL-AMS)

Example

```
entity OR2 is
          generic (DELAY_TIME       : TIME := 0 ns);
          port    (signal IN1, IN2  : in  BIT;
                   signal OUT1      : out BIT);
end entity OR2;
```

The entity declaration describes the interface of an OR gate. IN1 and IN2 are the names of the input ports. OUT1 is the name of the output port. DELAY_TIME is a constant that describes the delay between the change of values of inputs and outputs.

Note: rules for naming identifiers

In the example, OR2, DELAY_TIME, IN1, IN2, and OUT1 are all simple identifiers that are introduced by the user that developed the model. Please keep in mind:

- Identifiers must not be reserved words of the VHDL-AMS language. This is why, for instance, the name of the output port cannot be **out** because **out** is a reserved word. Reserved words are printed in boldface letters in code segments in this chapter. A full list of reserved word can be found in the 1076.1-1999 IEEE Standard VHDL Analog and Mixed-Signal Extensions.

- VHDL as well as VHDL-AMS are not case sensitive. Thus, for instance, delay_time, DELAY_TIME, and Delay_Time are all the same. This is also true for reserved words. For instance, **entity**, **Entity** and **ENTITY** are considered the same.
- Simple identifiers must start with an alphabetic letter followed by a letter, a digit, or an underline character ('_'). Identifiers cannot end with an underline character. Underline characters must be separated by a letter or a digit. Some examples of legal identifiers are:

```
legal_identifier
node23
InputVoltage
```

Design unit architecture

An **architecture** associated with an entity declaration describes the internal organization or operation of a design entity. An architecture describes the behavior, data flow, or structure of a design entity. It can be described using concurrent and simultaneous statements. Concurrent statements describe digital time-discrete behavior. Simultaneous statements describe analog time-continuous behavior. Concurrent statements are known from digital VHDL. Simultaneous statements are new language constructs in VHDL-AMS. Signals are the fundamental objects that carry digital waveforms. Concurrent statements update the values of signals. Quantities are the fundamental objects that carry analog waveforms. Simultaneous statements define relationships between quantities. However, quantities can be read in concurrent statements and signals can be read in simultaneous statements. Thus, not only digital and analog behavior can be expressed in VHDL-AMS, but the description of mixed-signal (analog-digital) behavior is also possible. The order of concurrent and simultaneous statements in an architectural body has no influence on the results of a simulation.

In conclusion, a design entity is an entity declaration together with an associated architecture body. Please keep in mind that a given entity declaration may be shared by many design entities, each of which has a different architecture.

Examples

```
architecture SIMPLE of OR2 is
begin
      OUT1 <= IN1 or IN2 after DELAY_TIME;
end architecture SIMPLE;
```

SIMPLE is an architecture that belongs to the entity declaration OR2. The output signal is determined by the input signals IN1 and IN2 using a concurrent statement. In a structural description the design entity which is characterized by the entity OR2 and the architecture SIMPLE is referenced as OR2(SIMPLE). It can be instantiated directly, for instance by

```
LABEL: entity OR2(SIMPLE)
         generic map (DELAY_TIME   => 2 ns)
          port map   (IN1  => ACTUAL_IN1,
                      IN2  => ACTUAL_IN2,
                      OUT1 => ACTUAL_OUT1);
```

The default value of DELAY_TIME is overwritten by the instance specification of 2 ns. The interface points are connected to the actual nets ACTUAL_IN1, ACTUAL_IN2, and ACTUAL_OUT1. In this example *named association* is used. The formal designators DELAY_TIME, IN1, IN2, and OUT1 that are used in the declaration of entity OR2 appear explicitly. In addition, *positional association* is possible. In this case, the formal designators do not appear in the parameter and/or port lists. An actual designator at a given position in an association list corresponds to the interface element at the same position in the interface list of the entity declaration. Thus, the following instantiation is also possible:

```
LABEL: entity OR2(SIMPLE)
       generic map (2 ns)
       port map    (ACTUAL_IN1, ACTUAL_IN2, ACTUAL_OUT1);
```

Design unit configuration

Structures can be described with the help of placeholders for the entity declarations. These placeholders are called components. The **configuration** unit is a construct that binds concrete models of building blocks (that is design entities) to placeholders in a structural description. This offers a very simple way to exchange design entities in a structural description. At present, configuration units are not supported in current implementations of some VHDL-AMS simulation engines. This is why we avoid the usage of configuration units in the following sections. The structural descriptions are mainly done using direct instantiation, which uses the design entity name directly.

Design unit package

A **package** declaration may contain the declarations of types, subprograms, files and so on. In this way, the reuse of description parts is supported. The declared elements can be made visible to other design units. The package STANDARD with declarations of standard types (for example BIT, BOOLEAN, REAL, ...) is part of the predefined language environment.

Design unit package body

A **package body** contains hidden parts of a package. For instance, the interface declaration of a subprogram is described in the package declaration. The package body contains the full program code. Its details are hidden outside the package.

6.2.2 Logical libraries and compilation of design units

Administration of analyzed design units

The design units are compiled into design libraries. The compilation consists of

- Analysis of the source code description
- Generation of an intermediary code which is saved in a design library

A *design library* must be created and managed using commands that depend on the specific simulation engine used. A physically existing design library is connected with a specific logical name. In VHDL-AMS descriptions, only the logical names are important and used. There are several kinds of design libraries. Compilation is usually done in the WORK library. Predefined libraries are available in the simulation environment and can be used as resource libraries. Examples are the STD library and the IEEE library.

WORK library

Compiled design units are by default placed into the working design library named WORK.

STD library

The STD library contains the standard packages STANDARD and TEXTIO. The packages predefine basic language types, subtypes, and functions. There are some extensions in the STANDARD package of VHDL-AMS compared to VHDL as follows

- An enumerated domain type is declared.

```
type DOMAIN_TYPE is (
     QUIESCENT_DOMAIN,    -- initialization phase
     TIME_DOMAIN,         -- transient analysis
     FREQUENCY_DOMAIN);   -- AC and noise
```

- A signal DOMAIN of type DOMAIN_TYPE declared in the standard package is set by the simulation engine and can be evaluated in the models.

```
signal DOMAIN : DOMAIN_TYPE := QUIESCENT_DOMAIN;
```

- The impure function NOW returns physical or real time. The result depends on the context in which the function is used.

```
subtype DELAY_LENGTH is TIME range 0 fs to TIME'HIGH;
impure function NOW return DELAY_LENGTH; -- phys. time
impure function NOW return REAL;         -- real time
```

- The predefined real array type is declared.

```
type REAL_VECTOR is array (NATURAL range <>) of REAL;
```

IEEE library

The IEEE library contains

- Packages where the designer finds standard logic system and the corresponding types and functions like STD_LOGIC_1164
- Packages with real-valued and complex-valued types, constants, and functions like MATH_REAL and MATH_COMPLEX respectively
- Packages with declarations of energy domains in VHDL-AMS like ELECTRICAL_SYSTEMS, MECHANICAL_SYSTEMS, and THERMAL_SYSTEMS [Std03]

EDA vendor specific libraries

In addition to standardized packages, EDA vendor specific libraries and packages exist. These packages often extend the functionality of standard packages. They can also be used in cases where standard packages are not available, or were not previously available. An example is the DISCIPLINES library available in the ADVance MS simulator of Mentor Graphics [MGC04] that summarizes declarations of energy domains such as ELECTROMAGNETIC_SYSTEM, KINEMATIC_SYSTEM, and so on. The library was available a long time before the IEEE finished standardization of the corresponding packages.

Reference of design units

The representation of a design unit in a design library is called a *library unit*. Library units can be made available in other design units by *context clauses*. These context clauses enable design units to be visible within other design units. Thus, design units can be referenced in other design units. A context clause is either a *library clause* or a *use clause*. A library clause defines logical library names that may be referenced in a design unit. The simple library clause looks like

```
library library_name;
```

After the reserved word **library** a list of library names may follow, such as:

```
library IEEE, DISCIPLINES;
```

A *use clause* achieves direct visibility of declarations. Each selected name in a use clause identifies one or more declarations that will potentially become directly visible.

```
use selected_name.item_name;
```

In addition, a list of selected names and item names can follow the reserved word **use**. If the *item*_name is the reserved word **all**, then the use clause identifies all declarations that are contained within the package or library denoted by the selected name.

Examples

```
use WORK.all;      -- comment starts with -- stops at the end of
line:
                   -- use clause makes all design units (for instance
                   -- all entity and architectural declarations)
                   -- from library WORK available

use IEEE.STD_LOGIC_1164.all;    -- makes all declarations from
                                -- the package STD_LOGIC_1164
                                -- from IEEE library available
```

The scope of the *use clause* starts immediately after the use clause. If the use clause occurs within the context clause of a design unit, the scope of the use clause extends to the end of the declarative region associated with the design unit. In practice all context clauses before an entity declaration are valid for the entity and all associated architectures. It is assumed that the following context clause is implicitly declared for each design unit. It is not necessary to add them to a VHDL-AMS description:

```
library STD, WORK;
use STD.STANDARD.all;
```

Rules and order of the analysis of design units

These organizational capabilities suggest complying with the following rules. Figure 6-1 demonstrates the procedure.

- Create text files with design units
  If possible follow the rule "One design unit – One design file". An exception is the description of the test-bench. This is the top level description. In this case the entity is empty. Entity declaration and architecture are generally saved in one file. Text files can be created with the preferred text editor.

- Compilation must be done in the correct order
  The rules defining the order in which the design units can be analyzed are direct consequences of the visibility rules. The VHDL-AMS standard (Section 11.4 of the standard) requires

  - A primary design unit whose name is referenced within a given design unit must be analyzed prior to the analysis of the given design unit.
  - A primary design unit must be analyzed prior to the analysis of any corresponding secondary design unit.

  If a design unit is changed, then all library units that are potentially affected become obsolete and must be reanalyzed before they can be used again. A primary design unit is affected by a change of a library unit where it is referenced. A library unit is a compiled design unit. A secondary design unit is potentially affected by a change in its corresponding primary unit. But a secondary unit does not affect the corresponding primary unit. For instance, a package body can be changed without recompilation of the corresponding package declaration.

- Simulation
  The simulation can begin after all design units used in a design have been compiled.

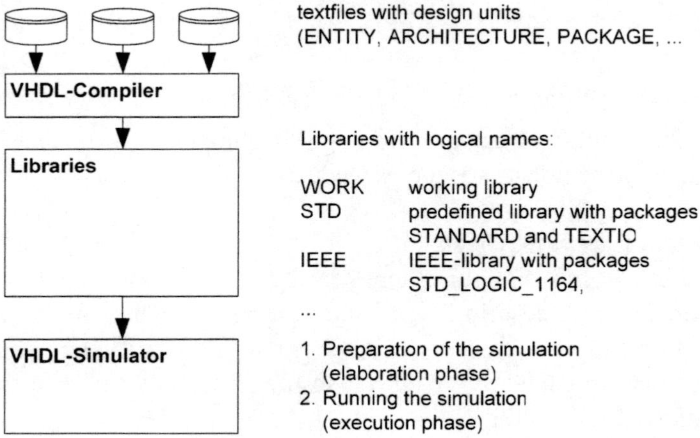

Figure 6-1. Analysis of design units

The commands to create a design library, compile a design unit and start the simulation engine are not part of the VHDL-AMS standard. They differ from one simulation engine to another one. For instance, in the simulation engine ADVance MS of Mentor Graphics [MGC04] the following commands are available in the command line of the operating system:

valib	to create a design library
vamap	to change the logical name of a design library
vacom	to compile a design unit into a design library
vasim	to invoke the simulation engine

6.2.3 Concurrent statements

Signals

A signal is an object with a history of past values. It can be considered a time-discrete waveform. Ranges and domains of such waveforms have two main properties:

- Values of signals may be of any type, for example BIT, BIT_VECTOR, and BOOLEAN. REAL and INTEGER value signals are also possible. Signals can also be user-defined types.
- The timeline of all signals is of type TIME. The resolution limit is the primary unit of type TIME. Any time value is a multiple of the resolution limit.

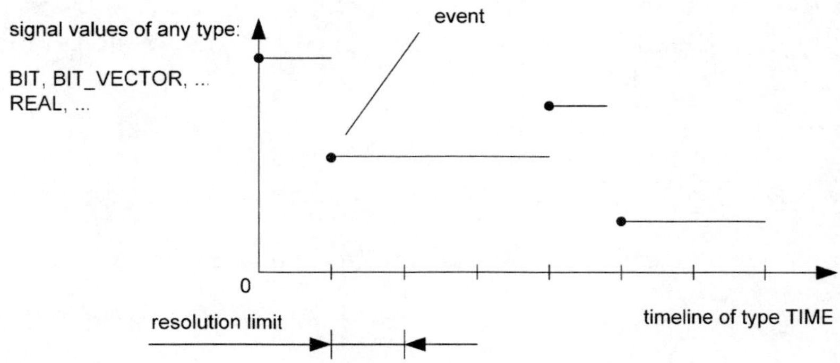

Figure 6-2. Signal in VHDL

The values of signals may only change at discrete time points as shown in Figure 6-2. A change in the signal value, which occurs when the signal is updated, is called an event. Signal values are constant between two events. Signals can be declared in the declaration part of an architecture. The initial value of a signal depends on the type (for example '0' for a BIT-valued signal) or can be overwritten during the declaration.

Examples of signal declarations

```
signal CLK       : BIT;
signal DATA1     : BIT_VECTOR (7 downto 0);
signal DATA2     : BIT_VECTOR (7 downto 0) := "00001111";
signal R_SIGNAL  : REAL := 1.0E4;            -- with initial value
```

Event-driven simulation

An appropriate algorithm to update signals is the event-driven simulation algorithm. The idea of event-driven simulation is to evaluate signals only at time points where a value change can occur. This procedure saves computation time by avoiding unnecessary signal evaluations. Concurrent statements determine the values of signals. Signal changes, which will occur in the future, are administrated using an event queue.

Simple concurrent signal assignment

Signal values can be changed by signal assignment statements. The concurrent statements are part of an architectural body. The simplest form of a concurrent signal assignment statement is

```
signal_name <= expression;
```

signal_name is the target of the concurrent statement. *expression* is the driver of the signal. A value may be assigned to a target signal after an explicit delay.

Examples

```
DATA1 <= "1010101010";
DATA2 <= "1111111111" after 10 ns;  -- with inertial delay
```

It is possible to apply multiple assignments in one statement:

```
R_SIGNAL <= 1.0E4, 1.0E3 after 10 ms, 1.0E6 after 100 ms;
```

The value of a signal can be determined using a mathematical expression. Operands can be signals, constants, and so on.

```
CLK <= not CLK after 1 ms;  -- CLK shall be of type BIT
```

The value of CLK changes from '0' to '1' after 1 ms and vice versa. Thus, the concurrent statement describes a simple clock generator.

Note: delay mechanisms in VHDL

There are two different delay mechanisms in VHDL. **inertial** delay is used by default. An existing transaction is always overwritten by a new transaction on the driver of a signal. For inertial delays, a new transaction scheduled after an existing transaction overwrites the existing transaction if it has a different value. If the delay is of type **transport**, the new transaction is appended to the event queue.

Example

The signals A and B drive C_INERTIAL_1_ns (A or B with inertial delay of 1 ns), C_INTERTIAL_2_ns (A or B with inertial delay of 2 ns), and C_TRANSPORT (A or B with transport delay of 2 ns). As a consequence of the delay mechanism a pulse that is smaller 2 ns is suppressed in C_INTERTIAL_2_ns but not in C_TRANSPORT.

```
entity BENCH is end entity BENCH;

architecture BENCH_DELAY of BENCH is
  signal A, B             : BIT;
  signal C_INERTIAL_1_ns  : BIT;
  signal C_INERTIAL_2_ns  : BIT;
  signal C_TRANSPORT      : BIT;
begin
  A <= '0', '1' after 10 ns, '0' after 11 ns;
  B <= '0', '1' after  2 ns, '0' after  5 ns, '1' after 15 ns;

  C_INERTIAL_1_ns <=           a or b after 1 ns;
  C_INERTIAL_2_ns <=           a or b after 2 ns;
  C_TRANSPORT     <= transport a or b after 2 ns;
end architecture BENCH_DELAY;
```

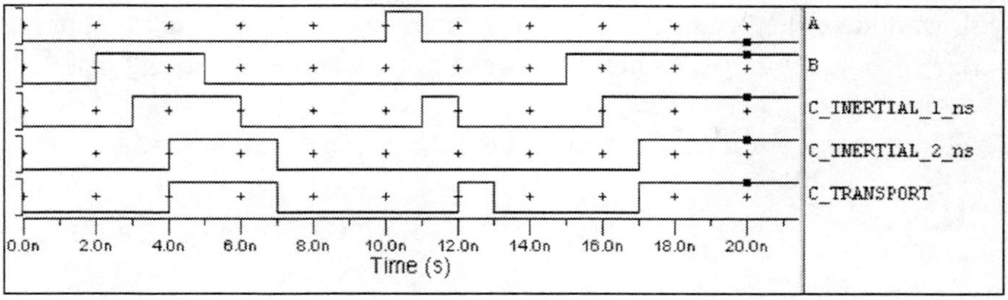

Figure 6-3. Results bench

Note: other concurrent statements

In addition to the simple concurrent signal assignment statement, more complex concurrent statements exist:

- Concurrent conditional signal assignment statement
- Concurrent selected signal assignment statement
- Concurrent procedure call statement
- Concurrent assertion statement
- Concurrent instantiation statement

They are introduced in Chapter 9 of the VHDL-AMS standard and will not be repeated in detail here. In the following we will only touch on the **process** statement.

Process statement

A special concurrent statement is the **process** statement. It allows defining the drivers of signals in a sequential way in the process statement part. The general form is

```
[process_label :]  process [ (sensitivity_list) ] is
process_declarative_part
begin
process_statement_part
end process [process_label];
```

A process is activated if an event occurs in one of the signals in the sensitivity list and runs until the end of the process is reached.

Processes without a sensitivity list must contain at least one **wait** statement. The wait statement causes the suspension of the process statements. The condition clause of a wait statement specifies a condition that must be met for the process to be executed. The execution of a process with wait statements consists of the repetitive execution of process statement part. A process with wait statements must not contain a sensitivity list.

The general form of the wait statement (VHDL-AMS standard Section 8.1) is

```
wait [ on    signal_name {, signal_name} ]
     [ until condition ]
     [ for   time_or_real_expression] ;
```

A process must contain either a sensitivity_list and no wait statement or no sensitivity_list and at least one wait statement.

Example

```
-- INPUT      is of type BIT_VECTOR (7 downto 0);
-- SIGNAL_OUT is a signal of type INTEGER
--            shall represent input as integer number

-- CLK        the conversion is done when an event
--            occurs on CLK

P1: process (CLK) is
  variable RESULT : INTEGER;
begin
  RESULT := 0;
  for I in 7 downto 0 loop
      if INPUT(I) = '1' then
         RESULT := 2 * RESULT + 1;
      end if;
  end loop;
  SIGNAL_OUT <= RESULT;
end process P1;
```

Process statements define the behavior in a sequential manner. In a process, variables can be declared. They can save intermediary results of an algorithm. The variables retain their value from one process call to the next. In contrast to a signal, the assignment of the value is performed immediately. The update of the signal value is not done until the next delta cycle of the event-driven simulation algorithm starts (see also Section 6.5.2).

Note: concurrent and sequential statements

In some cases the concurrent and sequential form of a statement are similar. In a process, only the sequential form of the statements can be applied.

- concurrent conditional statement

```
TARGET_SIGNAL <= 5.0 when clk = '1' else 0.0;
```

- sequential conditional statement

```
if CLK = '1' then
   TARGET_SIGNAL <= 5.0;
else
   TARGET_SIGNAL <= 0.0;
end if;
```

6.2.4 A simple pure digital example – divider

In this example we want to develop a VHDL model of a divider with the following requirements:

- Interface with input and output ports INP and OUTP respectively of type BIT
- Input signal frequency shall be divided by a positive integer number N

The model shall be tested with a test-bench with a BIT-valued clock signal that changes value after 1 ms. The clock signal is connected to the divider's input and shall be divided by N=5.

To follow the rules and order of the analysis of design units given in Section 6.2.2 we save entity, architecture and test-bench descriptions in different files:

```
-- File:     divider_ent.vhd
-- Content:  entity declaration of the DIVIDER
entity DIVIDER is
       generic (N     : POSITIVE);
       port    (INP   : in   BIT;
                OUTP  : out  BIT);
end entity DIVIDER;


-- File:     divider_simple.vhd
-- Content:  architecture SIMPLE of the DIVIDER
architecture SIMPLE of DIVIDER is
begin
  process is
    variable COUNTER : INTEGER := N-1;
  begin
    wait until INP = '1';  -- process with wait, no sens. list
    COUNTER := (COUNTER + 1) mod N;
    if COUNTER = 0 then
       OUTP <= '1';
    elsif COUNTER = N/2 then
       OUTP <= '0';
    end if;
  end process;
end architecture SIMPLE;

-- File:     bench.vhd
-- Contents: DIVIDER testbench
```

```
use WORK.all;  -- makes DIVIDER(SIMPLE) available

entity BENCH is end entity BENCH;

architecture BENCH_SIMPLE of BENCH is
  signal CLK, OUTP : BIT;
begin
  CLK <= not CLK after 1 ms;

UUT: entity DIVIDER(SIMPLE)
       generic map (5)           -- positional association
       port map     (CLK, OUTP);

end architecture BENCH_SIMPLE;
```

The compilation must begin with the file *divider_ent.vhd* (primary design unit on the lowest level of the model hierarchy). It is continued with the file *divider_simple.vhd* (secondary design unit of entity DIVIDER). At the end the file *bench.vhd* should be compiled. All files will be compiled into the logical library WORK. Thus, the context clause "use WORK.all;" was included in the test-bench description to allow direct instantiation of the design entity DIVIDER(SIMPLE). After five input pulses a new output pulse is created. The result is shown in Figure 6-4.

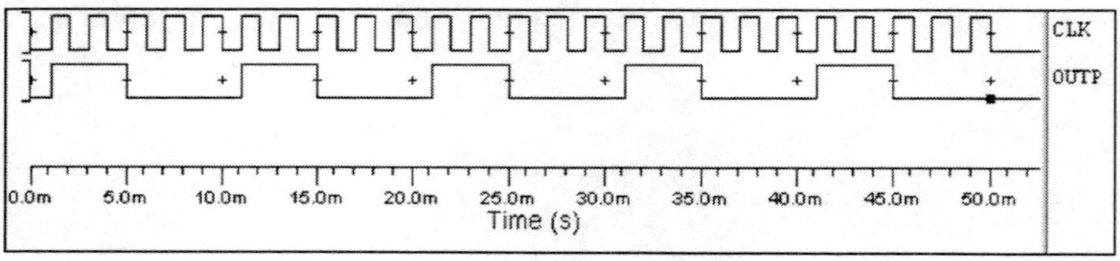

Figure 6-4. Results of divider simulation

6.3 Conservative Systems Description

Conservative semantics describe parts of the analog portion of a system. The modeled analog portions are similar to lumped systems, which can be described by ordinary differential equations and algebraic equations. Electrical networks are a special kind of conservative system.

The structure of a conservative system is characterized by the connection of its branches. The branches carry

- Across quantities (similar to branch voltages in the electrical case)
- Through quantities (similar to branch currents in the electrical case)

The following requirements have to be fulfilled by the solution of a conservative system

- Kirchhoff's Current Law (KCL)
- Kirchhoff's Voltage Law (KVL)
- Constitutive relations that describe requirements for branch voltages and currents

In this section, we will introduce some basic VHDL-AMS language constructs used to model conservative systems. Both electrical and nonelectrical systems can be modeled.

Background knowledge on network modeling approaches can be found in a wide range of books, such as [DeK69], [ChD87]. The description and simulation of the analog portion uses many ideas known from SPICE-like simulation engines. There are many books with more information (for example [Vla93], [VlS94], [Kun95]).

6.3.1 Network analysis problem

By means of an example we will look at the formulation of a network analysis problem and discuss the consequences concerning the language constructs required. The following circuit shall be considered.

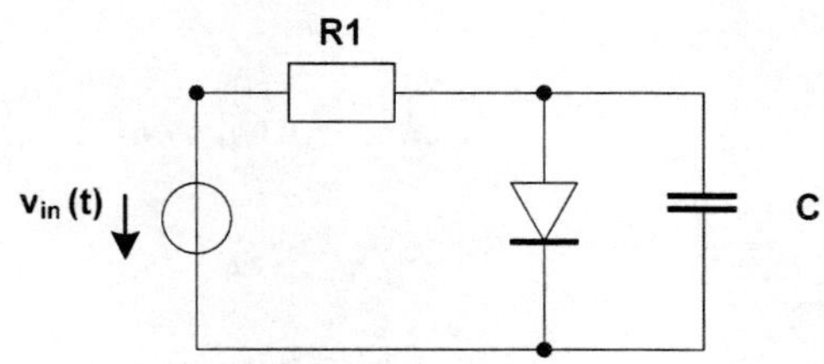

Figure 6-5. Network analysis problem

The input voltage *vin(t)* is given. The branch voltages and currents of all other branches shall be determined.

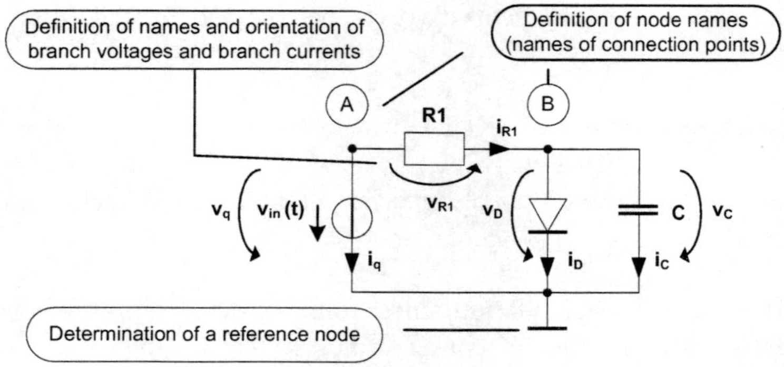

Figure 6-6. Schematic with node names and definition of branches

To establish the network equations we usually have to make some preparations:

- Definition of node names (names of connection points)
- Definition of names and orientation of branch voltages and branch currents
- Determination of a reference node

The structure of the circuit is given by its network graph. It describes how nodes are connected by oriented branches. Kirchhoff's Current and Voltage Law equations result from this graph.

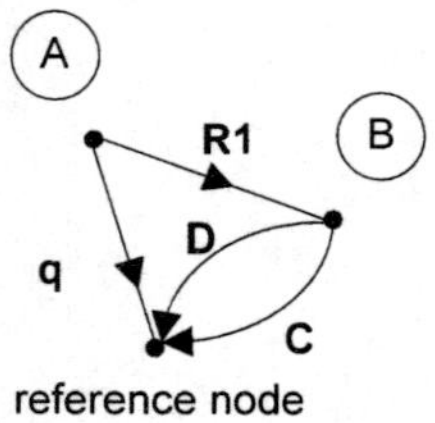

Kirchhoff's Current Law:

Node A => $i_q + i_{R1} = 0$

Node B => $-i_{R1} + i_D + i_C = 0$

Kirchhoff's Voltage Law:

$$v_{R1} + v_D - v_q = 0$$

$$v_C - v_D = 0$$

Figure 6-7. Network graph and equations resulting from Kirchhoff's laws

Kirchhoff's Current Law (KCL) requires that the sum of branch currents at a node with respect to their orientation equals zero. The equation for the reference node linearly depends on the equations for the other nodes. Thus, it will not be part of the system of network equations. *Kirchhoff's Voltage Law (KVL)* requires that the sum of branch voltages of a mesh with respect to their orientation equals zero. The equations which result from Kirchhoff's laws can be automatically established on the base of the network graph. That means they only depend on the network topology (see Figure 6-7 for the example).

Furthermore, the voltage-current constitutive relations of the branches must be fulfilled. These equations define further restrictions to branch voltages and currents (see Figure 6-8 for the example). They must be independent of the equations given by Kirchhoff's laws.

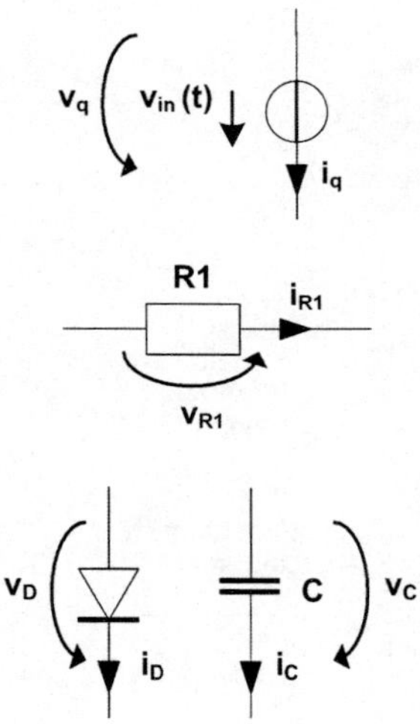

voltage source => $v_q - v_{in}(t) = 0$

resistor R1 => $v_{R1} = R1 \cdot i_{R1}$

diode => $i_D = f(v_{D,...})$

capacitor => $i_C = C \cdot \dfrac{dv_C}{dt}$

Figure 6-8. Symbols to describe the constitutive relations of branches

Conclusions

All these conditions (KCL, KVL, constitutive relations) must be fulfilled by the branch voltages and currents that solve the network analysis problem. A differential algebraic system of equations has to be evaluated:

$$F(x, \frac{dx}{dt}, p, t) = 0$$

with $x:[0,\infty) \to R^n$ (time $x \in [0,\infty)$ and fixed parameters $p \in R^m$)

The examples demonstrate the main tasks in modeling and simulating conservative systems:

- Objective of modeling
  - Description of network topology (network graph, that is node names, names and orientation of branch voltages and currents)
  - Description of the constitutive relation of the branches

- Task of the simulation engine
  - Elaboration of the model and establishing the system of network equations
  - Numeric solution of the network equations (see for example [LiZ97], Chapter 4)

What does the modeling language have to support?

The analog extensions of VHDL have to support the following modeling requirements for analog systems.

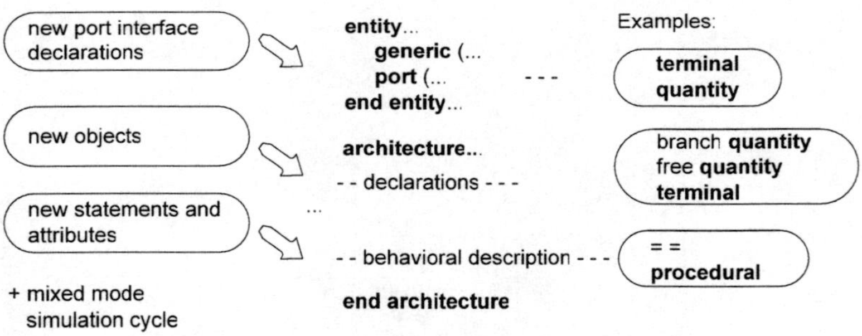

Figure 6-9. Requirements concerning analog extensions of VHDL

- New interface descriptions
  Beside signal ports it must be possible to describe *conservative connection points* (for example electrical pins, called **terminal** ports in VHDL-AMS) and, as will be shown later, also *nonconservative connection points* (for example signal flow pins of control blocks, called **quantity** ports in VHDL-AMS).

- New objects and data types
  The kind of a conservative connection point has to be declared. This can be done using a nature declaration in VHDL-AMS. The network branches can be described in VHDL-AMS using branch quantity declarations as will be shown later. Furthermore, additional unknowns

that may help to define the constitutive relations may be declared. They are called free quantities. Nodes with connected conservative interfaces points can also be declared. They are also called terminals.

- New statements
  The description of the constitutive relations can be done using *simultaneous statements*. VHDL-AMS has many facilities available to express these relations. Powerful attributes help to handle analog waveforms like 'DOT (for differentiation), 'INTEG (for integration), 'LTF (to describe Laplace transfer functions), and so on.

VHDL-AMS also defines how to exchange values between the analog and digital portions of a system during the simulation, which is the mixed-signal simulation cycle.

6.3.2 Nature, terminal and branch quantity declarations

Quantity

Analog waveforms in the time domain can be considered time-continuous waveforms.

- The range of values must be floating point types or subtypes.
- The timeline is of type REAL.
- Typical analog waveforms are branch across quantities, branch through quantities, and free quantities.

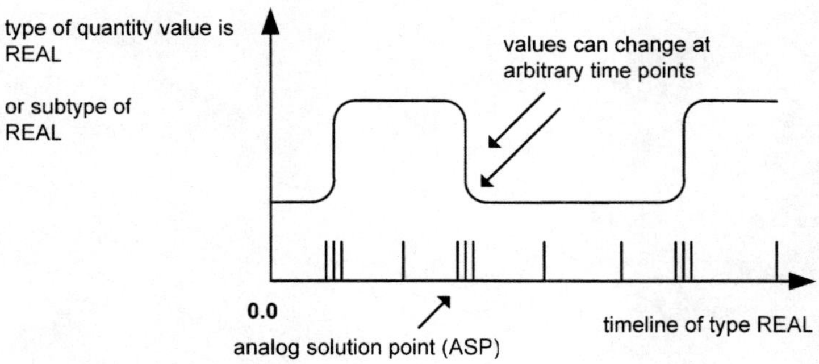

Figure 6-10. Analog waveform (quantity) in VHDL-AMS

The conditions concerning the analog waveforms are expressed by sets of so-called characteristic expressions. A consistent assignment of values to the quantities of a model is called the analog solution point (ASP), see Figure 6-10.

Nature declaration

A **nature** characterizes a node or a conservative interface connection point of an entity. Branches can only connect terminals of the same nature. The nature declaration specifies the types of the associated branch voltages and currents. That generally means across and through quantities. The declaration also includes the name of the associated reference terminal. The across quantity between a terminal of a given nature and its associated reference terminal is known as a "node voltage" in electrical networks. The type of this across quantity is the same type that was specified in the nature declaration. The general form is as follows:

```
nature nature_name  is
       across_type        across
       through_type       through
       reference_name     reference;
```

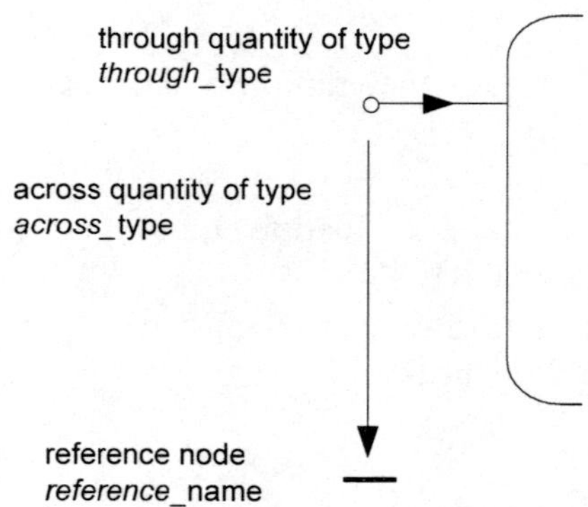

Figure 6-11. Elements of NATURE declaration

Example

```
nature ELECTRICAL is
     VOLTAGE            across
     CURRENT            through
     ELECTRICAL_REF     reference;
```

The nature declaration in this example is used in the IEEE package ELECTRICAL_SYSTEMS. VOLTAGE and CURRENT are subtypes of REAL. ELECTRICAL_REF can be used like a terminal without explicit declaration in an architecture. Natures may be of scalar and composite type. Composite natures are used to define a collection of terminals. They include arrays of terminals and records of terminals.

```
nature ELECTRICAL_VECTOR is
       array (NATURAL range <>) of electrical;
```

Packages with nature declarations

Nature declarations are usually collected in packages. In the IEEE library the ELECTRICAL_SYSTEMS package with the ELECTRICAL nature is given by the following code:

```
package ELECTRICAL_SYSTEMS is
   -- electrical domain
   -- subtype declarations
   subtype VOLTAGE     is REAL tolerance "DEFAULT_VOLTAGE";
   subtype CURRENT     is REAL tolerance "DEFAULT_CURRENT";
   -- ...
   -- nature declarations
   nature ELECTRICAL is
      VOLTAGE        across
      CURRENT        through
      ELECTRICAL_REF reference;
   nature ELECTRICAL_VECTOR is
          array (NATURAL range <>) of ELECTRICAL;
end package ELECTRICAL_SYSTEMS;
```

The reference node ELECTRICAL_REF of the nature ELECTRICAL means the same as node 0 in SPICE-like simulation engines. The reserved word **tolerance** in the subtype declarations provides a possibility of how to handle the accuracy of the associated quantities during the numerical simulation. The VHDL-AMS language does not give an exact definition of how to do this. Thus, presently it is handled in different manners in the simulation or simply ignored. To use the declarations of these packages a context clause has to be included in the VHDL-AMS descriptions. For instance, to use the nature ELECTRICAL and also ELECTRICAL_REF you have to add

```
library IEEE;
use IEEE.ELECTRICAL_SYSTEMS.all;
```

As it is clear that this context clause has to be included we often omit it in the following text.

Port terminal declaration

The general form of port declarations in an entity declaration is given by

```
port (port_interface_list);
```

An element of the port interface list, which describes conservative connection points of the nature *nature*_name, consists of

```
terminal port_terminal_name_list  : nature_name
```

Example

Let us have a look at the entity declaration of a resistor.

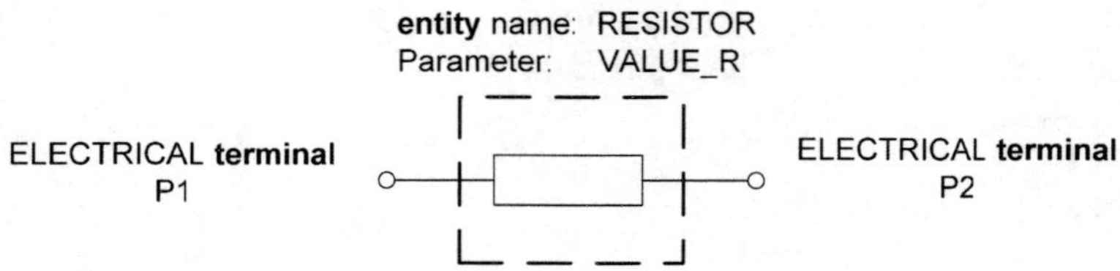

Figure 6-12. Interface description of a resistor

```
library IEEE;
    use IEEE.ELECTRICAL_SYSTEMS.all;

entity RESISTOR is
    generic (VALUE_R          : REAL := 1.0);
    port    (terminal P1, P2 : ELECTRICAL);
end entity RESISTOR;
```

Declaration of terminals for structural descriptions

Terminals can also be declared in an architecture body with a terminal declaration. From a network point of view they can be used as nodes. To connect port terminals in a structural description they have to be assigned to the same terminal (node). It is a similar situation as in digital VHDL where we declare signal ports and assign them to declared signals of an architecture. However, in VHDL-AMS the difference is that a terminal cannot carry any value. It is only the name of a connection point. The general form of a terminal declaration is

```
terminal node_name_list : nature_name;
```

Example

It is assumed that the entity declarations of a resistor and capacitor are compiled into the WORK library:

```
entity RESISTOR is
    generic (VALUE_R          : REAL := 1.0);
    port    (terminal P1, P2 : ELECTRICAL);
end entity RESISTOR;

entity CAPACITOR is
    generic (VALUE_C          : REAL := 1.0);
    port    (terminal P1, P2 : ELECTRICAL);
end entity CAPACITOR;
```

The V1 architectures V1 of the RESISTOR and CAPACITOR are also available in the WORK library. We show how to describe an RC chain in a hierarchical way. We start with the description of a simple RC subcircuit.

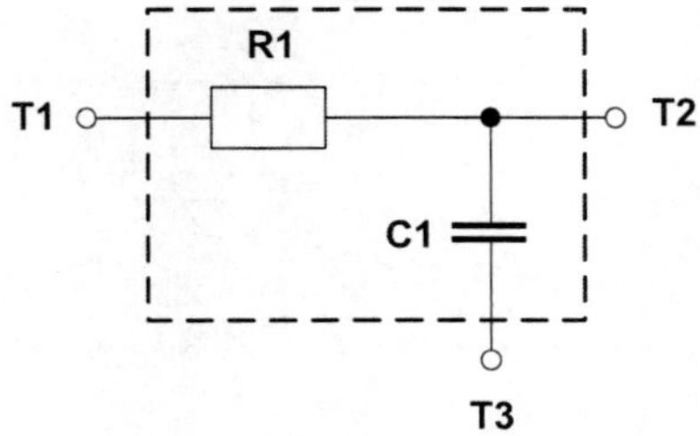

Figure 6-13. Simple RC subcircuit

```
library IEEE;
    use IEEE.ELECTRICAL_SYSTEMS.all;
    use WORK.all;

entity RC is
    generic   (RES              : REAL := 1.0;
               CAP              : REAL := 1.0);
    port      (terminal T1, T2, T3 : ELECTRICAL);
begin
    assert (RES > 0.0) and (CAP > 0.0)
      report "ERROR: RES and CAP must be > 0.0"
      severity ERROR;
end entity RC;

architecture V1 of RC is
begin

R1: entity RESISTOR(V1)  generic map (RES) port map (T1, T2);
C1: entity CAPACITOR(V1) generic map (CAP) port map (T2, T3);

end architecture V1;
```

The context clause has to be included in order to make available the nature ELECTRICAL and the design entities RESISTOR(V1) and CAPACITOR(V1). The design entities are directly instantiated.

The test-bench consists of a chain of two of these simple RC circuits connected to a voltage source. The interface of the voltage source is described by

```
entity STEP is
    generic (AMPL      : REAL := 1.0;
             T_DELAY   : TIME := 1 ms;
             T_RISE    : REAL := 1.0);
    port    (terminal P, N : ELECTRICAL);
end entity STEP;
```

The value of the voltage between P1 and P2 changes after time T_DELAY from 0 to AMPL with the rise time T_RISE. The top circuit description follows.

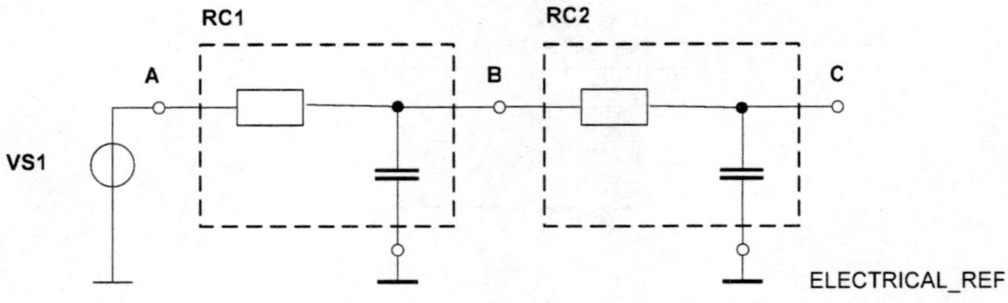

Figure 6-14. Test-bench for RC subcircuits

```
library IEEE;
    use IEEE.ELECTRICAL_SYSTEMS.all;
    use WORK.all;

entity BENCH is end entity BENCH;

architecture BENCH_RC of BENCH is
    terminal A, B, C : ELECTRICAL;
begin

VS1: entity STEP(V1) generic map (T_RISE => 2.0E-3)
     port map     (A, ELECTRICAL_REF);

RC1: entity RC(V1) generic map (4.0E2, 1.0E-6)
     port map     (A, B, ELECTRICAL_REF);

RC2: entity RC(V1) generic map (1.0E3, 1.0E-6)
     port map     (B, C, ELECTRICAL_REF);

end architecture BENCH_RC;
```

The simulation delivers the following results.

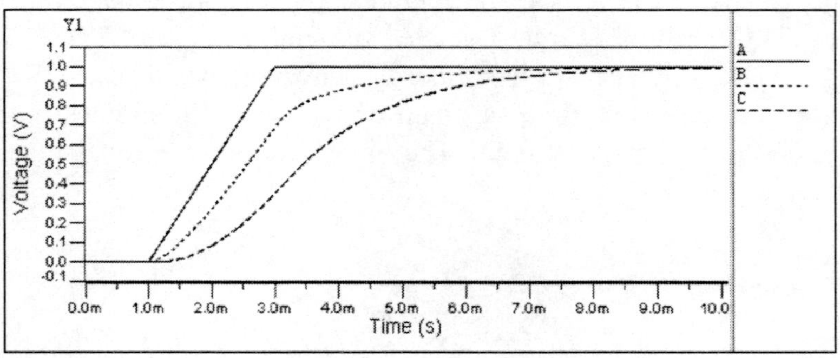

Figure 6-15. Simulation results

Branch quantity declaration

Branch quantities can be declared in an architecture body with a branch quantity declaration. The general form is

```
quantity [across_aspect] [through_aspect] terminal_aspect ;

across_aspect   ::= quantity_list  [:= expression] across
through_aspect  ::= quantity_list  [:= expression] through
terminal_aspect ::= start_terminal_name [ to end_terminal_name ]
```

A quantity named in an across aspect is an *across quantity*. Similarly, a quantity named in a through aspect is a *through quantity*. Both terminals of the terminal aspect must be of the same nature. The nature of the terminals determines the types of across and through aspects. Terminals may be port terminals of the associated entity of an architecture, or internally declared. If the terminal aspect only consists of a start terminal, the end terminal is the reference node of the nature of the start terminal. A branch quantity declaration terminals of an architecture.

If there is more than one quantity in the through aspect, parallel branches are declared must include at least an across or a through aspect. If a branch quantity declaration includes neither an across nor a through aspect it results in an error.

Notes

- A *branch quantity declaration without a through aspect* only declares a voltage between two terminals. A constitutive relationship must not be defined for a branch that is declared without a through aspect. In an electrical application a branch without a through aspect is an open branch.
- For each *branch quantity declaration with a through aspect* a constitutive relation has to be defined. In an electrical application that means a current can flow through such a branch.

Examples

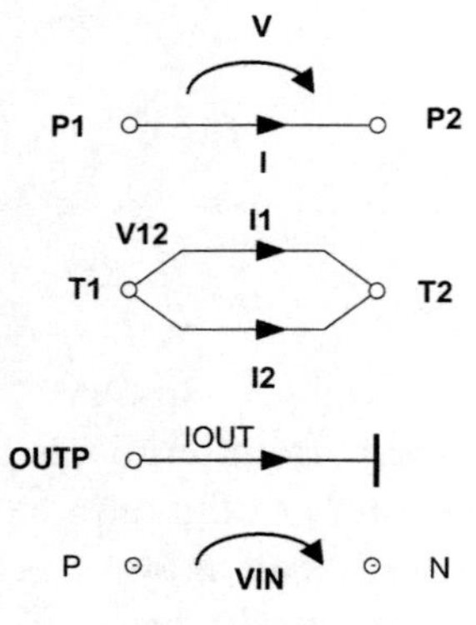


```
-- general form
quantity V across I through P1 to P2;

-- two parallel branches
quantity V12 across I1, I2 through T1 to T2;

-- branch ended with reference node
quantity VOUT across IOUT through OUTP;

-- open branch (branch current equal 0.0)
quantity VIN across P to N;
```

Figure 6-16. Branch quantity declarations (branch diagrams)

6.3.3 Simultaneous statements and free quantity declarations

Simultaneous statements express explicit and implicit differential and algebraic equations that constrain the values of the quantities of a model. In the case of conservative systems they describe the constitutive relations of network branches. The simultaneous statements must be placed in the architecture body as well as the concurrent statements. The order of simultaneous and concurrent statements does not matter.

General form:

[context_clause]

architecture *architecture*_name **of** *entity*_name **is**

declaration of internal terminals, branch quantities, ... ⇨ {declaration_part}

begin

description of the behavior, constitutive relations, ... ⇨ {simultaneous_statement | concurrent_statement}

end [architecture] [*architecture*_name];

Figure 6-17. Simultaneous statements in an architecture

Example

Let us look at a simple resistor. The interface is described by the entity declaration. In the architecture the internal branches and the constitutive relations of branches have to be described.

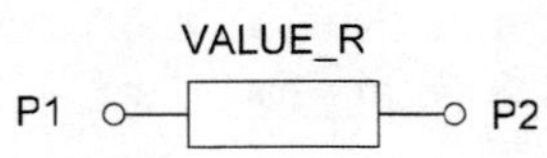


```
library IEEE;
  use IEEE.ELECTRICAL_SYSTEMS.all;

entity RESISTOR  is
  generic (VALUE_R        : REAL := 1.0);
  port    (terminal P1, P2  : ELECTRICAL);
end entity RESISTOR;

architecture V1 of RESISTOR is
```

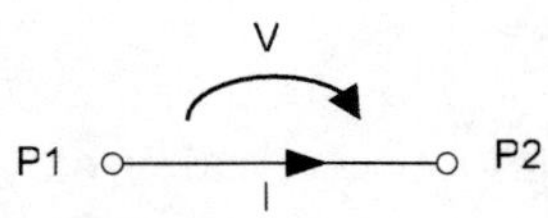

declaration of a branch from P1 to P2
branch voltage: V
branch current: I

```
begin
```

constitutive relation:
V = VALUE_R*I

simultaneous statement to describe
the constitutive relation

```
end architecture V1;
```

Figure 6-18. Structure of the resistor model

The internal branch is described by a branch quantity declaration. The voltage current relationship has to be expressed by a simultaneous statement. We show how this can be done in the following.

```
library IEEE;
  use IEEE.ELECTRICAL_SYSTEMS.all;

entity RESISTOR  is
  generic (VALUE_R        : REAL := 1.0);
  port    (terminal P1, P2  : ELECTRICAL);
end entity RESISTOR;

architecture V1 of RESISTOR is
```

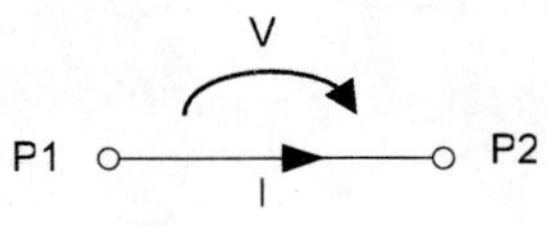


```
          quantity V across I through P1 to P2;

begin
          simultaneous statement to describe
          the constitutive relation

end architecture V1;
```

Figure 6-19. Resistor model with branch quantity declaration

Note

A simultaneous statement is required for each branch quantity declaration with a through aspect.

Simple simultaneous statement

The evaluation of a simple simultaneous statement creates a new characteristic expression that has to be taken into consideration during the solution of the network equations. The general form of a simple simultaneous statement is

```
[label:] simple_expression == simple_expression ;
```

The left-hand and right-hand side expressions must be real or of the same real subtype. Composite real types or subtypes are possible. The expressions are constructed using constants and quantities. The simulation engine determines the values of the quantities so that the difference of both expressions equals zero or is near to zero. That means, the simultaneous statement expresses a condition that has to be fulfilled. The equals sign == is not an assignment operator. The order of simultaneous and concurrent statements in an architecture does not influence the simulation results.

Example (resistor)

Now we can complete the architecture of the resistor.

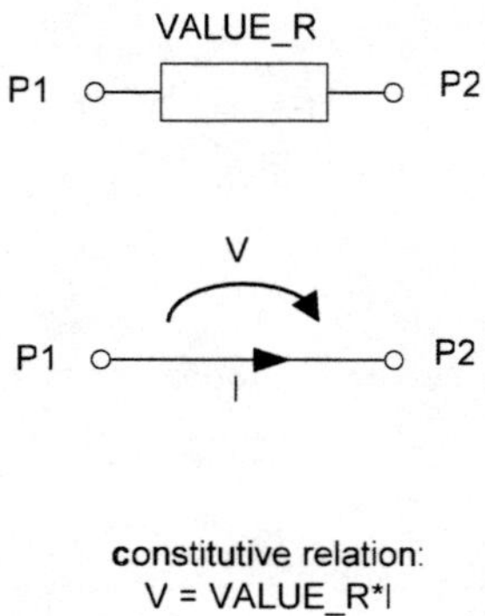


```
library IEEE;
  use IEEE.ELECTRICAL_SYSTEMS.all;

entity RESISTOR is
  generic (VALUE_R         : REAL := 1.0);
  port    (terminal P1, P2 : ELECTRICAL);
end entity RESISTOR;

architecture V1 of RESISTOR is
        quantity V across I through P1 to P2;
begin
        V == VALUE_R * I;
end architecture V1;
```

Figure 6-20. Complete resistor model

```
library IEEE;
    use IEEE.ELECTRICAL_SYSTEMS.all;

entity RESISTOR is
    generic (VALUE_R          : REAL := 1.0);
    port    (terminal P1, P2 : ELECTRICAL);
end entity RESISTOR;

architecture V1 of RESISTOR is
    quantity V across I through P1 to P2;
begin
    V == VALUE_R * I;
end architecture V1;
```

Example (sinusoidal voltage source)

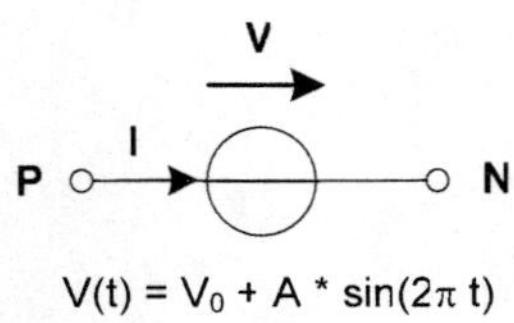

Figure 6-21. Sinusoidal voltage source

```
library IEEE;
   use IEEE.ELECTRICAL_SYSTEMS.all;
   use IEEE.MATH_REAL.all;        -- for access to SIN and MATH_2_PI


entity SINE is
    generic (V0   : REAL := 0.0;
             A    : REAL := 1.0;
             FREQ : REAL := 50.0);
    port (terminal P, N  : ELECTRICAL);
end entity SINE;

architecture V1 of SINE is
    quantity V across I through P to N;
begin
    V == V0 + A*SIN(MATH_2_PI*FREQ*NOW);
end architecture V1;
```

The function NOW delivers the current simulation time (t in the constitutive relation).

Example (voltage controlled voltage source)

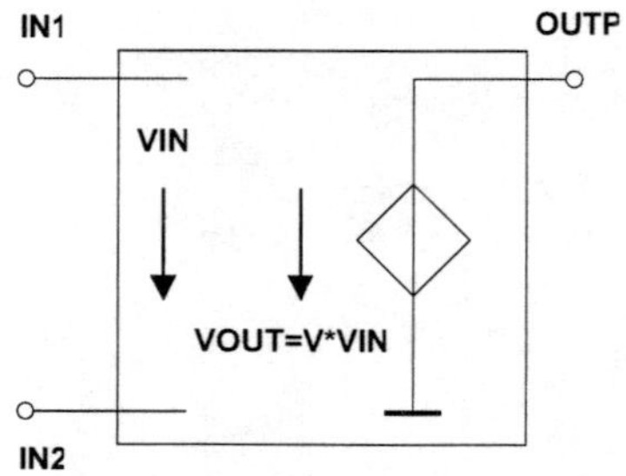

Figure 6-22. Voltage controlled voltage source

```
library IEEE;
   use IEEE.ELECTRICAL_SYSTEMS.all;

entity VCVS is
   generic (V      : REAL := 1.0);      -- gain
   port    (terminal IN1, IN2, OUTP : ELECTRICAL);
end entity VCVS;

architecture V1 of VCVS is
   quantity VIN  across IN1 to IN2;     -- open branch
   quantity VOUT across IOUT through OUTP;
begin
   VOUT == V*VIN;
end architecture V1;
```

The open input branch declaration is without a through aspect. The current in this branch is zero. Thus, there is only one branch declaration with a through aspect and one simultaneous statement in the architectural body.

Free quantity declaration

In the previous examples we expressed the constitutive relations with only the help of branch voltages and currents. In some cases we need auxiliary quantities to express the behavior. Therefore, we can declare so-called *free quantities* in an architecture. Each declared free quantity increases the number of required simultaneous statements. By default the initial value of the real-valued free quantity is 0.0. This value can be overwritten by a real-valued expression in the free quantity declaration. The general form of a free quantity declaration is

```
quantity name_list : real_type_or_subtype_name [ := expression] ;
```

Example

We add an input resistor and a series resistor to the output branch of the voltage controlled voltage source. The amplified input voltage equals the declared free quantity VCTRL. In a similar way, other blocks with input and output resistors can be modeled.

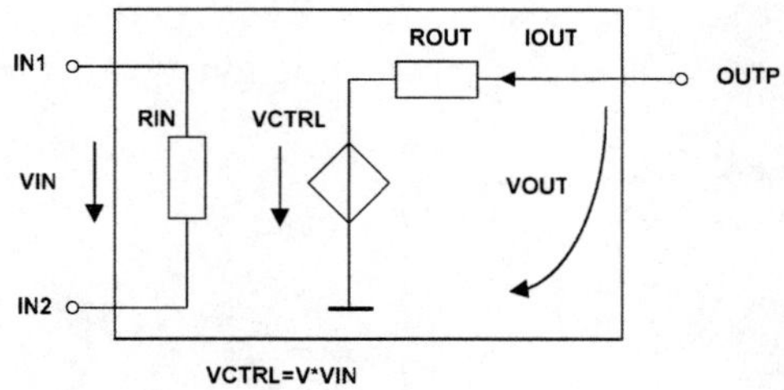

Figure 6-23. Voltage controlled voltage source with output resistor

```
library IEEE;
   use IEEE.ELECTRICAL_SYSTEMS.all;

entity VCVS_R is
   generic (V    : REAL :=  1.0;    -- gain
            RIN  : REAL := 50.0;    -- input resistor [Ohm]
            ROUT : REAL := 50.0);   -- output resistor [Ohm]
   port    (terminal IN1, IN2, OUTP : ELECTRICAL);
end entity VCVS_R;

architecture V1 of VCVS_R is
   quantity VIN  across IIN  through IN1 to IN2;    -- input branch
   quantity VOUT across IOUT through OUTP;          -- output branch
   quantity VCTRL : REAL;                        -- free quantity
begin
   VIN   == RIN*IIN;
   VCTRL == V*VIN;                          -- description of functionality
   VOUT  == VCTRL + ROUT*IOUT;
end architecture v1;
```

Further simultaneous statements

Simultaneous if statement

The simultaneous **if** statement selects one statement part for evaluation. The selection depends on the values of one or more conditions. Dynamic conditions are possible. That means the conditions may depend on values of signals and quantities that can change during the simulation. If one of the conditions evaluates to TRUE, then the corresponding simultaneous statement is evaluated. The general form is

```
[if_label :] if boolean_condition use
    simultaneous_statement_part
  { elsif boolean_condition use
    simultaneous_statement_part }
  [ else
    simultaneous_statement_part ]
  end use [if_label] ;
```

In a special case the selection is done between two simultaneous statements. Either the first or the second one must be evaluated.

```
if condition use
     simultaneous_statement_1
  else
     simultaneous_statement_2
end use ;
```

Simultaneous case statement

A simultaneous **case** statement selects one of a number of alternative statement parts for evaluation. The general form is

```
[case_label :] case expression use
      when choice { | choice } =>
               { simultaneous_statement }
    { when choice { | choice } =>
               { simultaneous_statement }  }
end case [case_label] ;
```

The expression must be a discrete type or a one-dimensional array type, whose element base type is a character type (for example BIT_VECTOR with a given length). Each value chosen must be the same type as the expression. The simple expression and discrete ranges specified as choices must be locally static. The choice **others** covers all values not specified in the choices of previous alternatives. It is only allowed for the last alternative.

An **others** choice is required in a case statement if the expression is a universal integer type (for example INTEGER), since this is the only way to cover all values of the universal integer type.

This is explained by the example below. The data object I is an integer valued expression. VOUT and VIN are branch quantities. KP, KI, and KD are real valued constants.

```
case I use
  when 1 => VOUT == KP*VIN;
  when 2 => VOUT == KI*VIN'INTEG;
  when 3 => VOUT == KD*VIN'DOT;
  when others => VOUT == KP*VIN + KI*VIN'INTEG + VIN'DOT;
end case;
```

Simultaneous procedural statement

The simultaneous **procedural** statement provides a sequential notation for expressing differential and algebraic equations. In the statement part, the sequential form of the statements must be applied in a similar way to their usage in a process statement. The procedural statement is a simultaneous equivalent to the concurrent process statement. The general form is

```
[procedural_label :] procedural [ is ]
      declaration_statement_part
begin
      sequential_statement_part
end procedural [ procedural_label ] ;
```

In the statement part, values can be assigned to quantities in a sequential order. Values of quantities can also be determined by a function. With the help of a simple simultaneous statement a quantity and the function value can be required to be equal. Thus, the simultaneous procedural statement always has an equivalent simple simultaneous statement. Nevertheless, it offers many advantages if similar expressions have to be computed in various simple simultaneous statements. This can be applied in transistor modeling for instance.

6.3.4 Example of a conservative system – A-law companding

To reduce the influence of noise in data transmission systems companding (compressing-expanding) is used. The waveform to be transmitted is compressed using a nonlinear amplitude characteristic [Kam92].

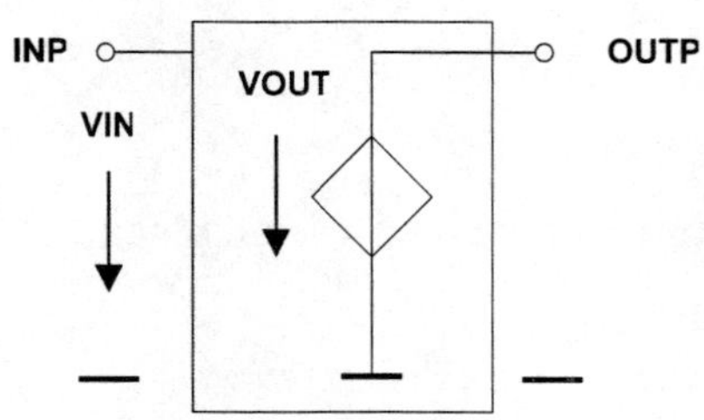

Figure 6-24. Interface of compression block

To reconstruct the waveform it must be expanded using the inverse characteristic. One scheme preferred in Europe is A-law companding. We model the block to compress the waveform in VHDL-AMS. The range of the input voltage VIN is between –VMAX and VMAX. The output voltage shall be given by the following formula. The ranges of input and output voltage shall be equal.

$$vout = \frac{A}{1+\ln A}\cdot\frac{vin}{v\max} \qquad \text{for } 0 < \left|\frac{vin}{v\max}\right| \le \frac{1}{A}$$

$$vout = v\max\cdot sign(\frac{vin}{v\max})\cdot\frac{1+\ln(A\cdot\left|\frac{vin}{v\max}\right|)}{1+\ln A} \qquad \text{for } \frac{1}{A} < \left|\frac{vin}{v\max}\right| < 1$$

Table 6-1 shows the coefficients of A.

Table 6-1. Coefficients of A

K	1	2	3	4	5	6
A	1.00	5.36	14.77	36.85	87.56	201.84

With A=1.0 a linear response is given. The commonly adopted value is A=87.56. We have to consider the following steps in the modeling procedure:

- Writing a VHDL-AMS for the compression block
- Generic parameters of the block shall be K and VMAX
- Check the model for K=5 and an input voltage of frequency F=1 kHz with

$$vin = 2.0\cdot\sin(2\pi\cdot f\cdot t)$$

Proposed solution

The entity description of the compression block is given by

```
library IEEE;
    use IEEE.ELECTRICAL_SYSTEMS.all;
    use IEEE.MATH_REAL.all;

entity COMPRESS is
    generic (K     : POSITIVE  := 1;   -- index for A array
             VMAX : REAL       := 1.0);
    port (terminal INP, OUTP : ELECTRICAL);
begin
    assert K <= 6
      report "ERROR: 1 <= K <= 6 required."
      severity ERROR;
```

```
     assert VMAX > 0.0
       report "ERROR: VMAX > 0.0 required."
       severity ERROR;
end entity COMPRESS;
```

The architecture is given by

```
architecture V1 of COMPRESS is
 quantity VIN  across INP;
 quantity VOUT across IOUT through OUTP;
 constant AK   : REAL_VECTOR (1 to 6)
       := (1.0, 5.36, 14.77, 36.85, 87.56, 201.84);
 constant A    : REAL   := Ak(k);
begin
 if ABS(VIN/VMAX) < 1.0/A use
  VOUT == A/(1.0+LOG(A))*vin;
 else
  VOUT == VMAX*SIGN(VIN/VMAX)*(1.0+LOG(A*ABS(VIN/VMAX)))
                                                   /(1.0+LOG(A));
 end use;
 assert VIN'ABOVE(-VMAX) and not VIN'ABOVE(VMAX)
   report "WARNING: VIN out of range."
   severity WARNING;
end architecture V1;
```

The 'ABOVE attribute which is used for dynamic range checking of VIN will be explained in Section 6.5.1. We will use the following description as a test-bench. An input branch is declared that connects N_IN and the reference node. The sinusoidal branch voltage is VIN and is defined by a simultaneous statement. The unit under test UUT is directly instantiated by a concurrent statement. Note that simultaneous and concurrent (for example instantiation) statements can be mixed in an architecture.

```
library IEEE;
    use IEEE.ELECTRICAL_SYSTEMS.all;
    use IEEE.MATH_REAL.all;
    use WORK.all;

entity BENCH is end entity BENCH;

architecture BENCH_A_LAW of BENCH is
    terminal N_IN, N_OUT : ELECTRICAL;
    quantity VIN across IIN through N_IN;
    constant AMPL : REAL := 2.0;
    constant FREQ : REAL := 1.0E3;
begin
    VIN == AMPL*SIN (MATH_2_PI*FREQ*NOW);

UUT: entity COMPRESS(V1)
       generic map (K => 5, VMAX => AMPL)
       port map    (INP => N_IN, OUTP => N_OUT);

end architecture BENCH_A_LAW;
```

Figure 6-25 shows the result of the simulation (node voltages at N_IN and N_OUT).

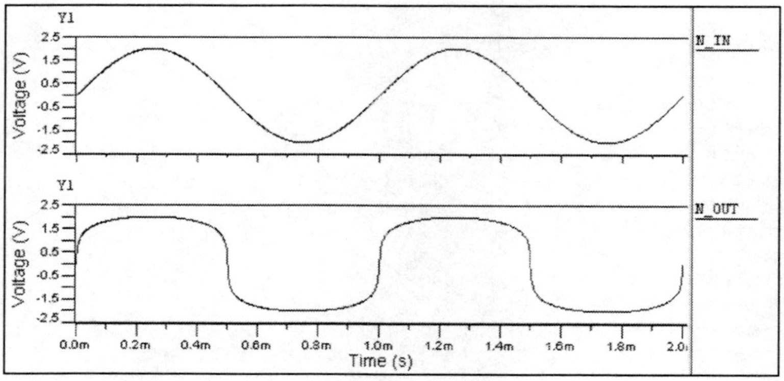

Figure 6-25. Input and output voltages versus time

We can represent the output voltage versus input voltage and obtain the A-law characteristic for A=87.56.

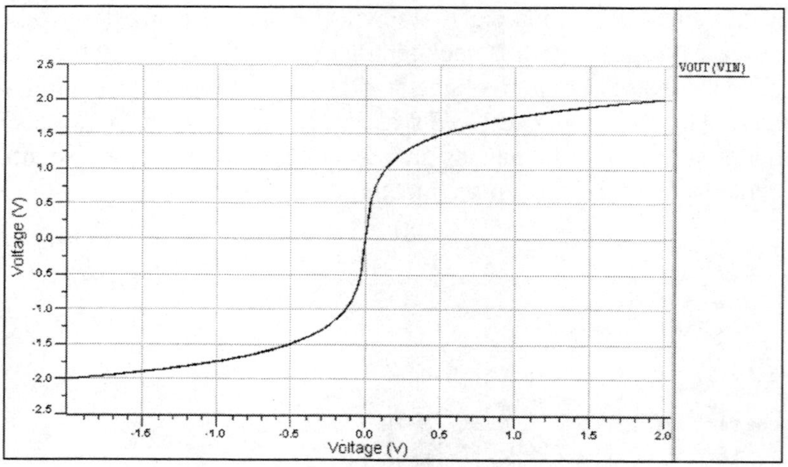

Figure 6-26. A-law characteristic

6.3.5 Attributes in VHDL-AMS

What is an attribute in VHDL-AMS?

An attribute is a definition of some characteristic of a named object. Some attributes are predefined for types, ranges, values, signals, quantities, and functions. A predefined attribute may return a constant value, a type, or a range. In some other cases it can create a new implicit signal or quantity. Many attributes are known from digital VHDL.

Table 6-2. Attributes in VHDL

Attribute name	Prefix	Result
T'LEFT	T is scalar type	Left bound of T
T'RIGHT	T is scalar type	Right bound of T
T'HIGH	T is scalar type	Upper bound of T
T'lLOW	T is scalar type	Lower bound of T
A'LENGTH (N)	A is an array	Length of the Nth index range. N=1 is omitted
A'LENGTH	A is an array (one dimensional)	Length of the first index range
A'LEFT	A is an array (one dimensional)	Left bound of the index range
A'RIGHT	A is an array (one dimensional)	Right bound of the index range
A'RANGE	A is an array (one dimensional)	Index range of A

New important attributes on quantities to describe the analog behavior are explained in the following:

- 'DOT to derive a quantity
- 'INTEG to integrate a quantity
- 'SLEW to smooth a quantity
- 'DELAYED to delay a quantity
- 'LTF to describe an analog filter
- 'ZOH to sample and hold a quantity
- 'ZTF to describe a digital filter

Other new attributes are introduced to describe mixed-signal behavior (see Section 6.5.1). The remaining attributes are user-defined and always constant. User-defined attributes are not taken into consideration.

Attribute 'DOT

The 'DOT attribute is characterized in the following way:

- Q'DOT is a quantity that is the derivative with respect to time of quantity Q at the time the attribute is evaluated.
- Q'DOT is an implicit quantity. It must not be declared.
- By default during the quiescent domain analysis (DC analysis) Q'DOT is zero.
- At a discontinuity by default Q is continuous if Q'DOT is used somewhere in a model.

Example

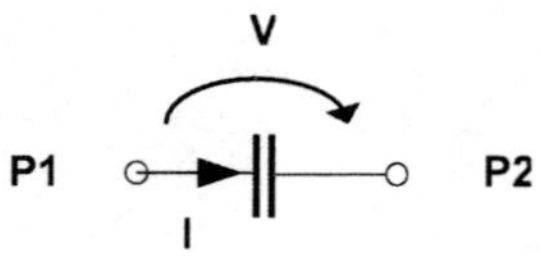

$$I = VALUE\_C \cdot \frac{dV}{dt}$$

Figure 6-27. Capacitance

```
library IEEE;
    use IEEE.ELECTRICAL_SYSTEMS.all;

entity CAPACITOR is
    generic (VALUE_C           : REAL := 1.0);
    port    (terminal P1, P2 : ELECTRICAL);
end entity CAPACITOR;

architecture V1 of CAPACITOR is
    quantity V across I through P1 to P2;
begin
    I == VALUE_C * V'DOT;
end architecture V1;
```

Attribute 'INTEG

The 'INTEG attribute is characterized in the following way:

- Q'INTEG is a quantity that is the time integral of quantity Q from time 0 to the time the attribute is evaluated.
- Q'INTEG is an implicit quantity. It must not be declared.
- By default during the quiescent domain analysis (DC analysis) Q is zero if Q'INTEG is used somewhere in the model.
- At a discontinuity by default Q'INTEG is continuous.

Example

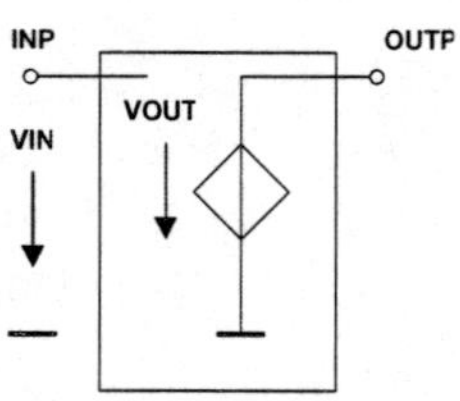

$$VOUT(t) = \int_0^t VIN(\tau) + IC$$

Figure 6-28. Integrator with electrical terminals

```
library IEEE;
    use IEEE.ELECTRICAL_SYSTEMS.all;

entity INTEGRAL_BLOCK is
  generic (IC     : REAL := 0.0);    -- initial value
  port (terminal INP  : ELECTRICAL;
        terminal OUTP : ELECTRICAL);
end entity INTEGRAL_BLOCK;

architecture IDEAL of INTEGRAL_BLOCK is
  quantity VIN  across INP;                 -- open input branch
  quantity VOUT across IOUT through OUTP; -- voltage source
begin
  if DOMAIN = QUIESCENT_DOMAIN use
     VOUT == IC;                -- during DC analysis
  else
     VOUT == VIN'INTEG + IC;  -- in transient analysis
  end use;
end architecture IDEAL;
```

The DOMAIN signal depends on the actual status of the simulation. During operating point (DC) analysis its value is QUIESCENT_DOMAIN.

Attribute 'SLEW

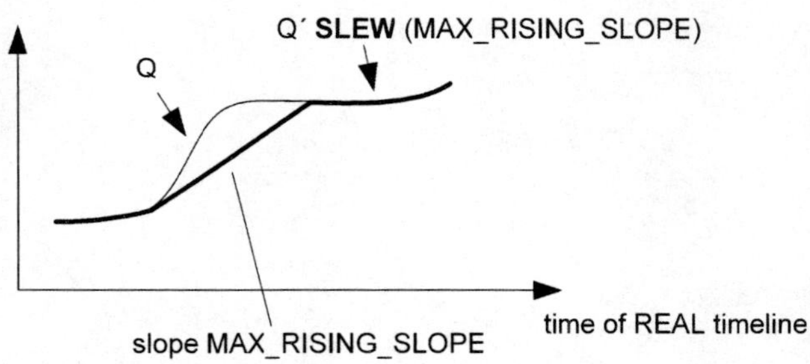

Figure 6-29. Quantities Q and Q'SLEW(MAX_RISING_SLOPE)

The 'SLEW attribute is characterized in the following way:

- Suppose Q is a scalar or composite quantity.
- Q'SLEW (MAX_RISING_SLOPE, MAX_FALLING_SLOPE) is a quantity where each scalar subelement follows the corresponding scalar subelement of Q, but its derivative with respect to time is limited by specified slopes.
- MAX_RISING_SLOPE is a static expression of type REAL that evaluates to a positive value. If omitted it defaults to REAL'HIGH, which is interpreted as an infinite slope. MAX_FALLING_SLOPE is a static expression of type REAL that evaluates to a negative value. If omitted it

defaults to the negative of MAX_RISING_SLOPE. The value REAL'LOW is interpreted as a negative infinite slope.

- Q'SLEW is an implicit quantity. It must not be declared.
- The derivative of Q'SLEW is between MAX_FALLING_SLOPE and MAX_RISING_SLOPE, that is

```
MAX_FALLING_SLOPE <= Q'SLEW (MAX_R…,MAX_F…)'DOT <= MAX_RISING_SLOPE
```

Q'SLEW follows Q as long as Q'DOT is between MAX_FALLING_SLOPE and MAX_RISING_SLOPE

Example

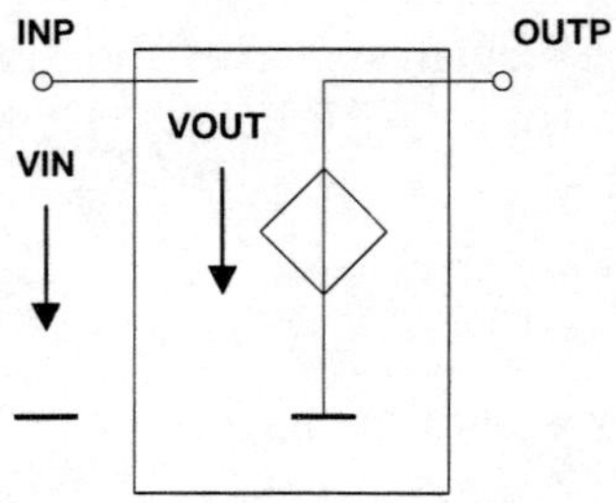

$$VOUT(t) = \max_{\tau<t} VIN(\tau)$$

Figure 6-30. Peak detector [CoC92]

```
library IEEE;
     use IEEE.ELECTRICAL_SYSTEMS.all;

entity PEAKDETECTOR is
    port (terminal INP, OUTP : ELECTRICAL);
end entity PEAKDETECTOR;

architecture IDEAL of PEAKDETECTOR is
    quantity VIN  across              INP;
    quantity VOUT across IOUT through OUTP;
begin
    VOUT == VIN'SLEW(REAL'HIGH, -1.0e-38);
end architecture IDEAL;
```

VOUT follows VIN if VOUT is increasing. Otherwise it retains the last value of VIN. MAX_FALLING_SLOPE value of -1.0E-38 is similar to 0.0.

Input voltage source

$$VIN(t) = AMP \cdot e^{A \cdot t} \cdot \sin(2\pi \cdot FREQ \cdot t)$$

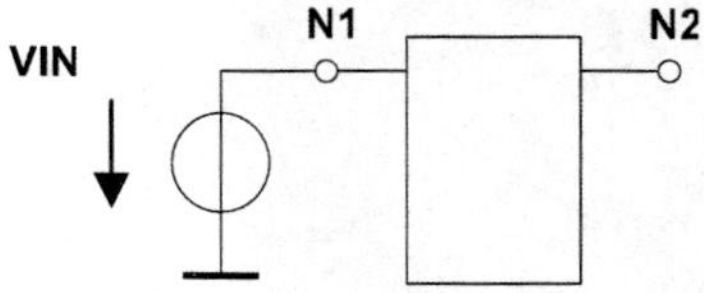

Figure 6-31. Test-bench for peak detector

```
library IEEE;
    use IEEE.ELECTRICAL_SYSTEMS.all;
    use IEEE.MATH_REAL.all, WORK.all;

entity BENCH is end entity BENCH;

architecture BENCH_PEAKDETECTOR of BENCH is
    terminal N1, N2       : ELECTRICAL;
    quantity VIN across IIN through N1;
    constant AMP  : REAL := 1.0;
    constant A    : REAL := 1.0e3;
    constant FREQ : REAL := 1.0e3;
begin
    VIN == AMP*EXP(A*NOW)*SIN(MATH_2_PI*FREQ*NOW);

UUT: entity PEAKDETECTOR(IDEAL)
       port map (INP => N1, OUTP => N2);
end architecture BENCH_PEAKDETECTOR;
```

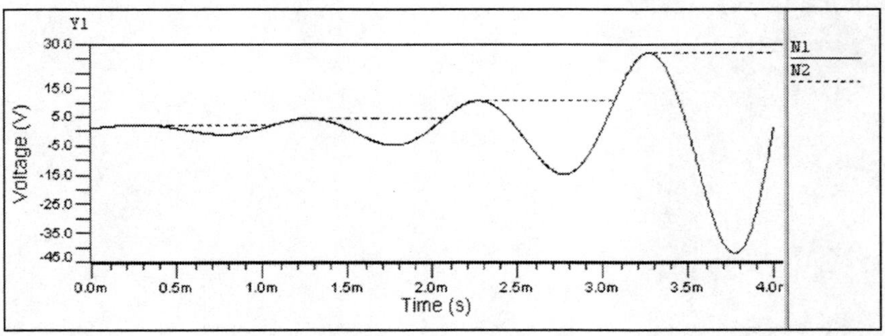

Figure 6-32. Simulation results of peak detector

Attribute 'DELAYED

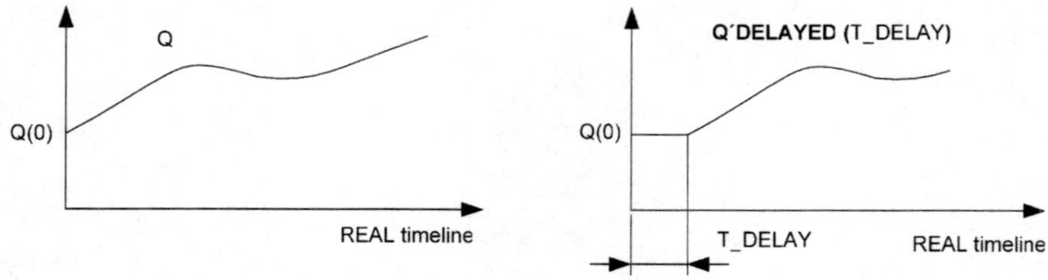

Figure 6-33. Quantity Q and delayed waveform Q'DELAYED(T_DELAY)

The 'DELAYED attribute is characterized in the following way:

- Q'DELAYED(T_DELAY) is a quantity equal to quantity Q delayed by T_DELAY. T_DELAY is a static expression of type REAL that evaluates to a non-negative number. If omitted it defaults to 0.0.
- Q'DELAYED is an implicit quantity. It must not be declared.
- During DC analysis (DOMAIN equals QUIESCENT_DOMAIN) Q'DELAYED equals Q.
- Between time 0 and time T_DELAY the value of Q'DELAYED(T_DELAY) equals the value of Q at time 0.

Example (delay block)

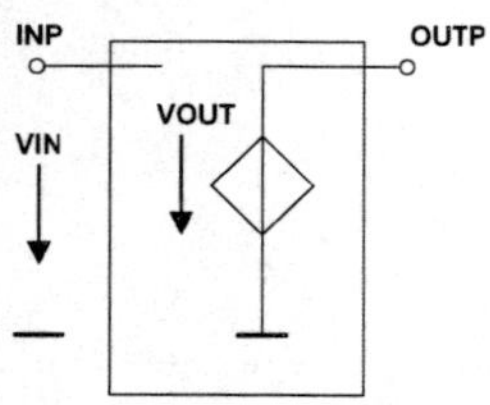

VOUT(t) = VIN(t-T_DELAY)

Figure 6-34. Delay block

```
library IEEE;
    use IEEE.ELECTRICAL_SYSTEMS.all;

entity DELAY_BLOCK is
    generic (T_DELAY : REAL);  -- delay time [s]
    port (terminal INP, OUTP : ELECTRICAL);
begin
    assert T_DELAY >= 0.0
      report "T_DELAY must be >= 0.0"  severity ERROR;
```

```
end entity DELAY_BLOCK;

architecture IDEAL of DELAY_BLOCK is
    quantity VIN  across               INP;
    quantity VOUT across IOUT through OUTP;
begin
    VOUT == VIN'DELAYED(T_DELAY);
end architecture IDEAL;
```

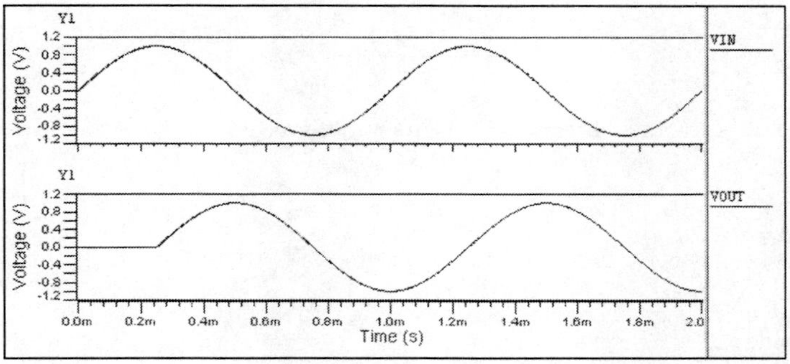

Figure 6-35. Examples for input and output voltages VIN and VOUT respectively

Example (lossless line)

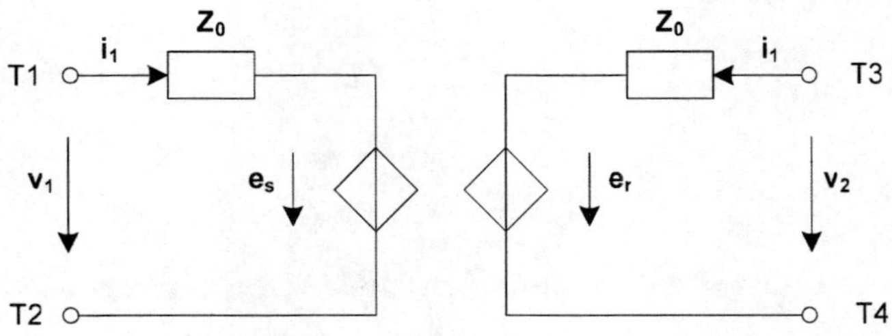

Figure 6-36. Network model of lossless line

The lossless line model is based on Branin's approach [Bra67]. Parameters of the model are the length l of line, the inductance L' per unit length, and the capacitance C' per unit length. The parameters determine the wave resistance Z_0 and the delay time T: $Z_0 = \sqrt{\frac{L'}{C'}}$ and $TD = \sqrt{L' \cdot C'} \cdot l$ respectively. Then the following equations describe the line model:

$$v_1(t) = Z_0 \cdot i_1(t) + e_s(t), \qquad e_r(t) = 2 \cdot v_1(t - TD) - e_s(t - TD),$$

$v_2(t) = Z_0 \cdot i_2(t) + e_r(t)$, $e_s(t) = 2 \cdot v_2(t - TD) - e_r(t - TD)$.

The VHDL-AMS model implements these equations using the 'DELAYED attribute.

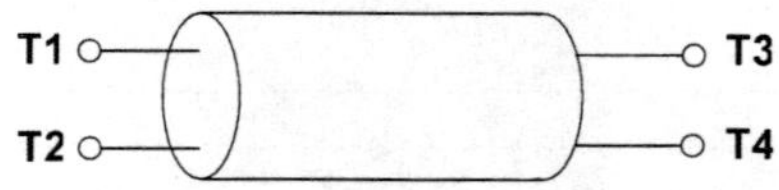

Figure 6-37. Terminals of lossless line

```
library IEEE;
    use IEEE.ELECTRICAL_SYSTEMS.all;

entity LINE is
    generic (Z0 : REAL := 50.0;        -- wave resistance [Ohm]
             TD : REAL := 1.0e-3);     -- delay time [s]
    port (terminal T1, T2, T3, T4 : ELECTRICAL);
end entity LINE;


architecture LOSSLESS of LINE is
    quantity V1 across I1 through T1 to T2;
    quantity V2 across I2 through T3 to T4;
    quantity ER, ES      : REAL;
begin
    V1 == Z0*I1 + ES;
    V2 == Z0*I2 + ER;
    ER == 2.0*V1'DELAYED(TD)  - ES'DELAYED(TD);
    ES == 2.0*V2'DELAYED(TD)  - ER'DELAYED(TD);
end architecture LOSSLESS;
```

The model is tested with a simple circuit. A 1 ms pulse is used as input.

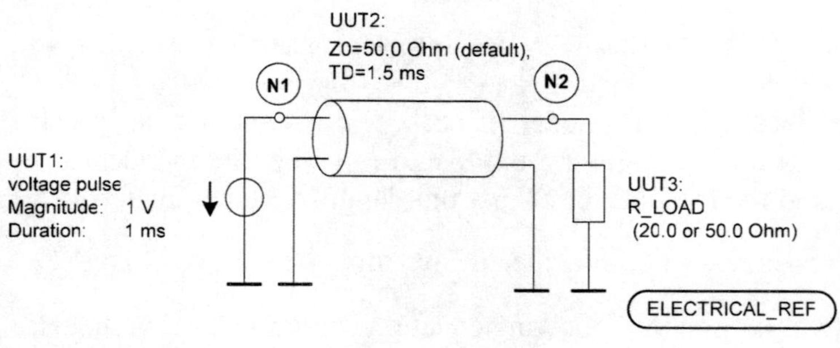

Figure 6-38. Test-bench for lossless line model

```
library IEEE;
    use IEEE.ELECTRICAL_SYSTEMS.all;
    use WORK.all;

entity BENCH is end entity BENCH;

architecture BENCH_LOSSLESS_LINE of BENCH is
    constant R_LOAD       : REAL := 20.0;
    terminal N1, N2       : ELECTRICAL;
begin

V1: entity V_SOURCE(PULSE)
       generic map (DURATION => 1 ms)
       port map (P => N1, N => ELECTRICAL_REF);

UUT: entity LINE(LOSSLESS)
       generic map (TD => 1.5e-3)
       port map (T1 => N1, T2 => ELECTRICAL_REF,
                 T3 => N2, T4 => ELECTRICAL_REF);

R1: entity RESISTOR(V1)
       generic map (VALUE_R => R_LOAD)
       port map (N2, ELECTRICAL_REF);

end architecture BENCH_LOSSLESS_LINE;
```

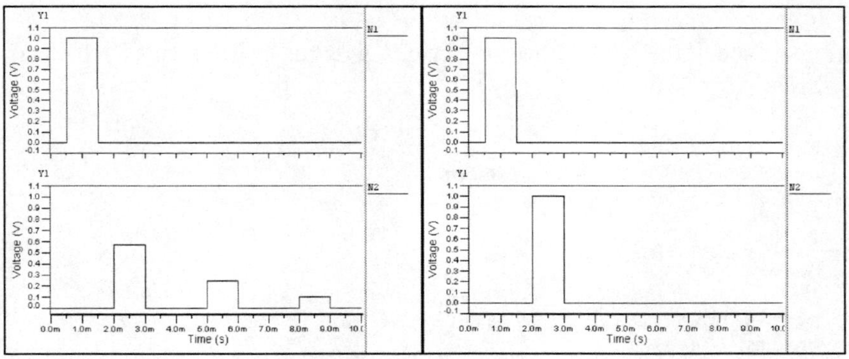

Figure 6-39. Results for R_LOAD=20 Ω and R_LOAD=50 Ω

Attribute 'LTF

The 'LTF attribute is characterized in the following way:

- Q'LTF (NUM, DEN) is a quantity that results in the application of Laplace transfer function on a quantity Q.
- Assume the Laplace transfer function is given by

$$H(s) = \frac{a_0 + a_1 \cdot s + a_2 \cdot s^2 + ... + a_m \cdot s^m}{b_0 + b_1 \cdot s + b_2 \cdot s^2 + ... + b_n \cdot s^n}$$

- NUM is a static expression of type REAL_VECTOR that contains the numerator coefficients, that is NUM equals (a0, a1, a2, ..., am).
- DEN is a static expression of type REAL_VECTOR that contains the denominator coefficients, that is DEN equals (b0, b1, b2, ..., bm). The first scalar subelement of DEN must not be 0.0.
- Q'LTF is an implicit quantity. It must not be declared.

Example

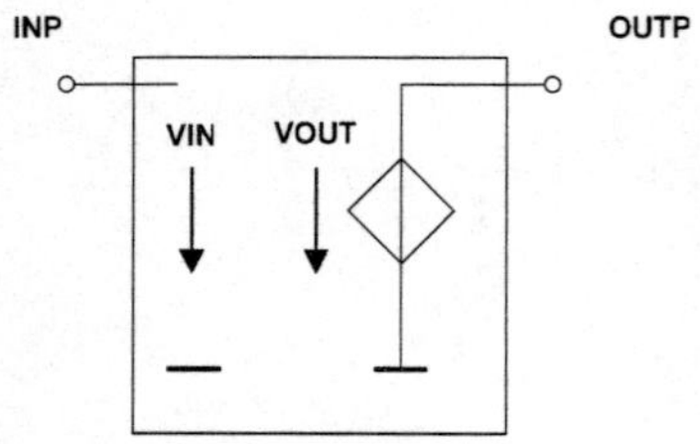

Laplace transfer function

$$H(s) = \frac{gain}{1 + \frac{1.3617 \cdot s}{2\pi \cdot f_c} + \frac{0.618 \cdot s^2}{(2\pi \cdot f_c)^2}}$$

Figure 6-40. Interface of lowpass model

The Laplace transfer function describes a second order Bessel lowpass filter with cut-off frequency f_c [TiS02].

```
library IEEE;
    use IEEE.ELECTRICAL_SYSTEMS.all;
    use IEEE.MATH_REAL.all;

entity LOWPASS is
    generic (FC                 : REAL;
             GAIN               : REAL := 1.0);
    port (terminal INP, OUTP : ELECTRICAL);
end entity LOWPASS;

architecture BESSEL_2 of LOWPASS is
    constant W   : REAL := MATH_2_PI*FC;
    constant NUM : REAL_VECTOR := (0 => 1.0);
    constant DEN : REAL_VECTOR
             := (1.0, 1.3617/W, 0.6180/W/W);
    quantity VIN  across              INP;
    quantity VOUT across IOUT through OUTP;
begin
    VOUT == GAIN*VIN'LTF(NUM,DEN);
end architecture BESSEL_2;
```

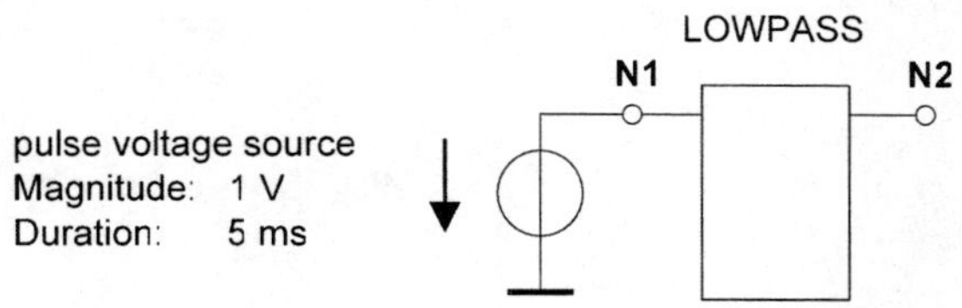

Figure 6-41. Test-bench of lowpass model (5 ms input pulse)

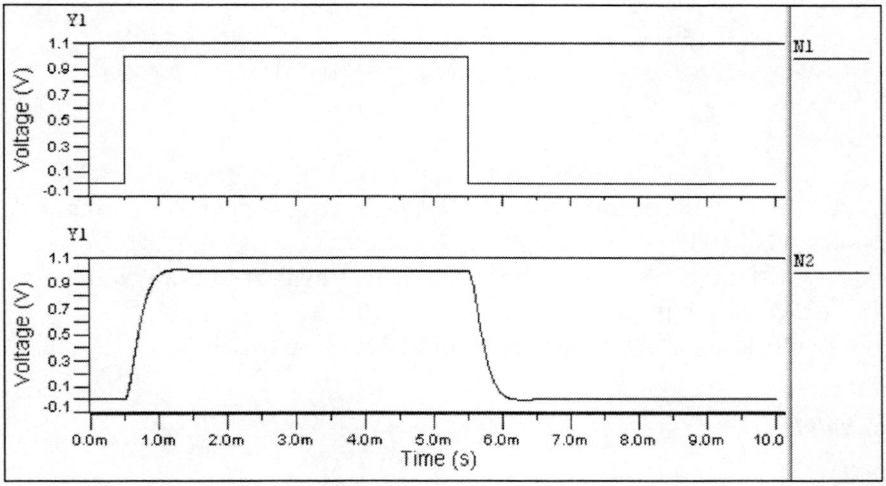

Figure 6-42. Results of test-bench simulation (Bessel lowpass filter, $f_c = 1$ kHz)

Attribute 'ZOH

The 'ZOH attribute is characterized in the following way:

- Q'ZOH (T, INITIAL_DELAY) is a quantity where the value of each scalar subelement is set to the value of the corresponding scalar subelement of Q at the sampling times INITIAL_DELAY + k×T (where k is any non-negative integer) and held constant until the next sampling time.

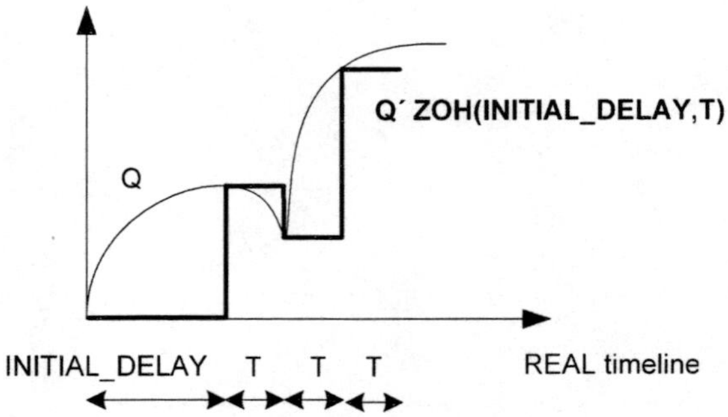

Figure 6-43. Quantity Q and waveform Q'ZOH (INITIAL_DELAY, T)

- INITIAL_DELAY is a static expression of type REAL. The first sampling will occur after INITIAL_DELAY seconds. If omitted it defaults to 0.0.
- T is a static expression of type REAL that evaluates to a positive value. This is the sampling period.
- Q'ZOH is an implicit quantity. It must not be declared.

Example

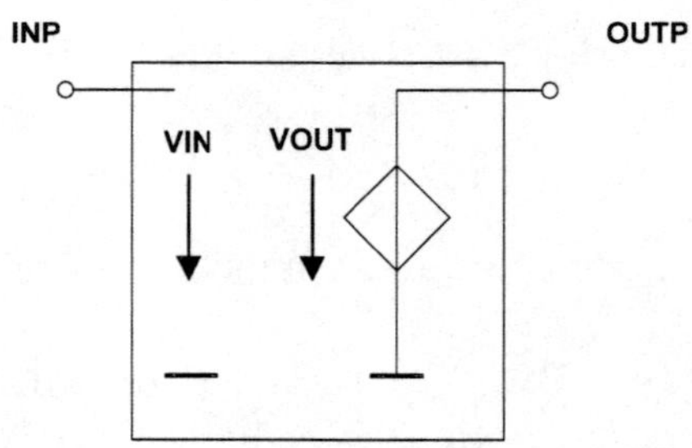

Figure 6-44. Sample and hold block

```
library IEEE;
    use IEEE.ELECTRICAL_SYSTEMS.all;

entity SAMPLE_AND_HOLD is
    generic (TSAMPLE   : REAL := 1.0E-3;
             DELAY     : REAL := 0.0);
    port (terminal INP, OUTP : ELECTRICAL);
end entity SAMPLE_AND_HOLD;
```

```
architecture IDEAL of SAMPLE_AND_HOLD is
    quantity VIN  across             INP;
    quantity VOUT across IOUT through OUTP;
begin
    VOUT == VIN'ZOH(TSAMPLE, DELAY);
end architecture IDEAL;
```

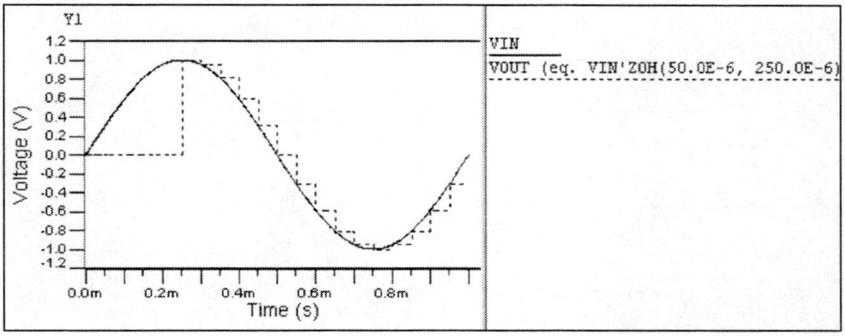

Figure 6-45. Results of a test
(1 kHz sinusoidal input voltage sampled after 250 μs with a period of 50 μs)

Attribute 'ZTF

The 'ZTF attribute is characterized in the following way:

- Q'ZTF (NUM, DEN, T, INITIAL_DELAY) is a quantity that results in the application of a z-domain transfer function on a quantity Q.
- Assume the z-domain transfer function is given by

$$H(z) = \frac{a_0 + a_1 \cdot z + a_2 \cdot z^{-2} + ... + a_m \cdot z^{-m}}{b_0 + b_1 \cdot z + b_2 \cdot z^{-2} + ... + b_n \cdot z^{-n}}$$

- NUM is a static expression of type REAL_VECTOR that contains the numerator coefficients, that is NUM equals (a0, a1, a2, …, am).
- DEN is a static expression of type REAL_VECTOR that contains the denominator coefficients, that is DEN equals (b0, b1, b2, …, bm). The first scalar subelement of DEN must not be 0.0.
- T is the sampling period and INITIAL_DELAY is the time of the first sampling. If omitted it defaults to 0.0.
- Q'ZTF is an implicit quantity. It must not be declared.

Example

The z-domain transfer function of a digital second order Bessel lowpass filter with cut-off frequency f_c and sampling frequency $f_s = 4\,f_c$ is given [Sch92].

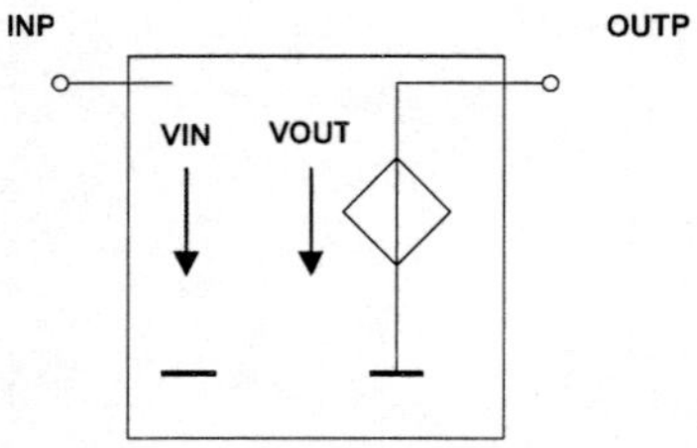

z-domain transfer function

$$H(z) = gain \cdot \frac{0.3356 + 0.6712 \cdot z^{-1} + 0.3356 \cdot z^{-2}}{1 + 0.2564 \cdot z^{-1} + 0.0860 \cdot z^{-2}}$$

Figure 6-46. Interface of digital lowpass model

```
library IEEE;
    use IEEE.ELECTRICAL_SYSTEMS.all;
    use IEEE.MATH_REAL.all;

entity LOWPASS is
    generic (FC                     : REAL;
             GAIN                   : REAL := 1.0);
    port (terminal INP, OUTP : ELECTRICAL);
end entity LOWPASS;

architecture BESSEL_2_DIGITAL of LOWPASS is
    constant FS      : REAL := 4.0*FC;   -- only in this case
    constant TSAMPLE : REAL := 1.0/FS;
    constant NUM : REAL_VECTOR
             := (0.3356, 0.6712, 0.3356);
    constant DEN : REAL_VECTOR
             := (1.0000, 0.2564, 0.0860);
    quantity VIN  across              INP;
    quantity VOUT across IOUT through OUTP;
begin
    VOUT == GAIN*VIN'ZTF(NUM,DEN, TSAMPLE, 0.0);
end architecture BESSEL_2_DIGITAL;
```

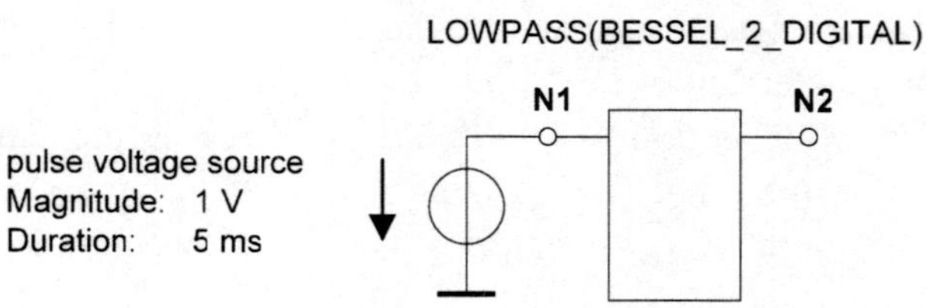

Figure 6-47. Test-bench of digital lowpass model (5 ms input pulse)

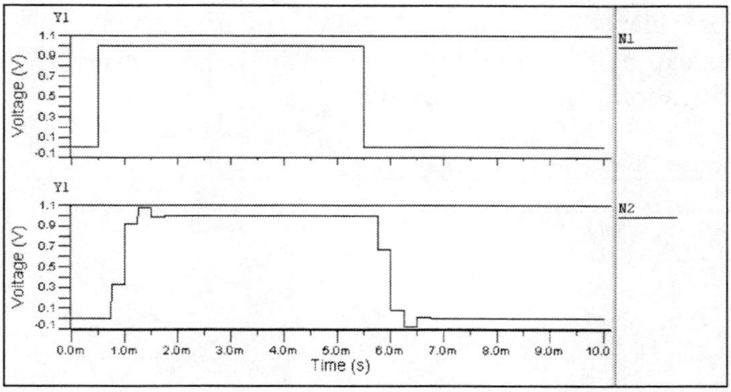

Figure 6-48. Results of test-bench simulation (digital Bessel lowpass filter)

Further Attributes

Table 6-3 summarizes additional new attributes of VHDL-AMS to describe analog behavior in detail.

Table 6-3. Further VHDL-AMS attributes

Attribute name	Prefix	Result
N'ACROSS	N is any nature	Across type of the nature denoted by N
N'THROUGH	N is any nature	Through type of the nature denoted by N
T'REFERENCE	T is any terminal	Across quantity whose plus terminal is T and whose minus terminal is the reference terminal of the nature of T ("node voltage")
T'CONTRIBUTION	T is any terminal	Contribution quantity of terminal T
T'TOLERANCE	T is any floating point type or subtype T	String with the tolerance group of T
Q'TOLERANCE	Q is any scalar quantity denoted by the static name Q	String with the tolerance group of Q

6.3.6 Example – higher order lowpass filter

The 'LTF attribute, mentioned in the previous section, provides a very simple mechanism in VHDL-AMS to describe filter functions using their coefficients. The transfer functions of lowpass filters are often described as a product of second order filters

$$H(S) = \frac{1}{\prod_i (1 + c_i \cdot S + d_i \cdot S^2)} \quad \text{where } S = \frac{s}{2\pi \cdot f_c}$$

The coefficients are specified by tables of filter coefficients. A higher order filter can be expressed by a combination of lower order filters using free quantities for intermediate quantities.

Table 6-4. Bessel filter coefficients [TiS02]

order	i	c_i	d_i
1	1	1	0.0
2	1	1.3617	0.6180
3	1	0.7650	0.0
3	2	0.9996	0.4772
4	1	1.3397	0.4889
4	2	0.7743	0.3890

Looking at the table, the transfer function of a 4th order Bessel lowpass filter is, for instance using $\omega = 2\pi \cdot f_c$, given by

$$H(s) = \frac{gain}{1 + 1.3397 \cdot \frac{s}{\omega} + 0.4889 \cdot \frac{s^2}{\omega^2}} \cdot \frac{1}{1 + 0.7743 \cdot \frac{s}{\omega} + 0.3890 \cdot \frac{s^2}{\omega^2}} = H_1(s) \cdot H_2(s)$$

A free quantity Q1 is declared in the VHDL-AMS model. Q1 results from filtering the input voltage VIN by $H_1(s)$. The output voltage VOUT equals Q1 filtered by $H_2(s)$. In a similar way other filters of a higher order can be described. The model for the Bessel lowpass filter follows:

```
library IEEE;
    use IEEE.ELECTRICAL_SYSTEMS.all;
    use IEEE.MATH_REAL.all;

entity LOWPASS is
    generic (FC     : REAL ;
             GAIN   : REAL := 1.0);
    port (terminal INP, OUTP : ELECTRICAL);
end entity LOWPASS;

architecture BESSEL_4 of LOWPASS is
    quantity VIN  across              INP;
    quantity VOUT across IOUT through OUTP;
    quantity Q1   : REAL;

    constant W    : REAL         := MATH_2_PI*FC;
    constant NUM1 : REAL_VECTOR := (0 => GAIN);
    constant DEN1 : REAL_VECTOR
                  := (1.0, 1.3397/W, 0.4889/W/W);
    constant NUM2 : REAL_VECTOR := (0 => 1.0);
    constant DEN2 : REAL_VECTOR
                  := (1.0, 0.7743/W, 0.3890/W/W);
begin
    Q1   == VIN'LTF(NUM1, DEN1);
    VOUT == VIN'LTF(NUM2, DEN2);
end architecture BESSEL_4;
```

6.4 Description of Nonconservative Systems

Nonconservative semantics describe parts of analog systems where Kirchhoff's laws do not apply. The energy in nonconservative systems does not remain constant, but it is added from or lost to the outside of the system.

In modeling we usually name those systems nonconservative, where unidirectional signals instead of through and across quantities are present, for instance when modeling control systems. Ports of nonconservative terminals carry only analog waveforms. Control subsystems can be modeled using nonconservative ports. These ports are characterized in the following way:

- Nonconservative ports are so-called quantity ports. They also carry a direction mode **in** or **out**.
- The general form of the interface description of nonconservative
  input ports look like
  ```
  quantity  identifier_list : in real_type
  ```
  The mode **in** can be omitted.
  output ports look like
  ```
  quantity  identifier_list : out real_type
  ```
- If a quantity interface element within an interface list includes a default expression as for example
  ```
  quantity port_identifier : in real_typ := expression;
  ```
  if the port is unassociated the value of the input quantity equals the expression.
- The identifiers of the quantity ports can be used in the associated architecture just like quantities.
- Each quantity port of mode **out** increases the number of required simultaneous statements in the associated architecture.

In a structural description

- Quantity ports can be associated to quantities.
- A quantity port of mode **in** may be unconnected or unassociated only if its declaration includes a default expression.
- A quantity port of mode **out** may be unconnected if its type is not an unconstrained array.
- If any quantity is associated as an actual with more than one formal of mode **out** an error results.

It should be checked whether the simulation tool being used supports unassociated ports.

Example

An ideal proportional plus integral plus derivative controller with input Q_IN and output Q_OUT is described by the following equation:

$$Q\_OUT(t) = KR \cdot \left(Q\_IN(t) + \frac{1}{TN} \int_0^t Q\_IN(\tau)d\tau + TV \cdot \frac{dQ\_IN}{dt}(t) \right)$$

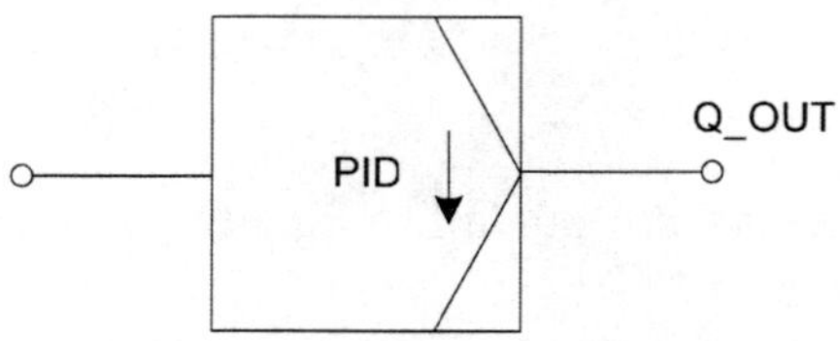

Figure 6-49. Interface of a PID controller

The architecture IDEAL of the entity PID realizes this functionality. The initial value of the integral is zero and not considered during operating point analysis (DOMAIN is QUIESCENT_DOMAIN).

```
entity PID is
  generic (KR : REAL := 1.0;    -- controller gain
           TN : REAL := 1.0;    -- reset time
           TV : REAL := 0.0);   -- derivative time
  port (quantity Q_IN  : in  REAL;   -- input quantity
        quantity Q_OUT : out REAL);  -- output quantity
begin
  assert TN /= 0.0
    report "ERROR: Reset time unequal 0.0 required."
    severity ERROR;
end entity PID;

architecture IDEAL of PID is
begin
  if DOMAIN = QUIESCENT_DOMAIN use
     Q_OUT == KR*(Q_IN + TV*Q_IN'DOT);
  else
     Q_OUT == KR*(Q_IN + 1.0/TN*Q_IN'INTEG + TV*Q_IN'DOT);
  end use;
end architecture IDEAL;
```

An instantiation of the model is shown in the following listing. The design entity QPWL(BASIC) provides the controller with the input waveform which is determined by the time value pairs that build up the parameter WAVE. Both models used were compiled into the WORK library. The free quantities INPUT and OUTPUT carry analog waveforms and are used as connections to the nonconservative ports of the PID controller.

```
use WORK.all;

entity BENCH is end entity BENCH;

 architecture BENCH_PID of BENCH is
    quantity INPUT, OUTPUT : REAL;   -- connection points
 begin
 Q1: entity QPWL(BASIC)
        generic map (WAVE =>
            (0.0, 0.0, 1.0, 1.0, 2.0, 1.0, 4.0, -1.0, 5.0, -1.0,
             6.0, 0.0, 10.0, 0.0))
        port map (INPUT);

 UUT: entity PID(IDEAL)
        generic map (KR => 5.0, TN => 2.0, TV => 1.0)
        port map (Q_IN => INPUT, Q_OUT => OUTPUT);
 end architecture BENCH_PID;
```

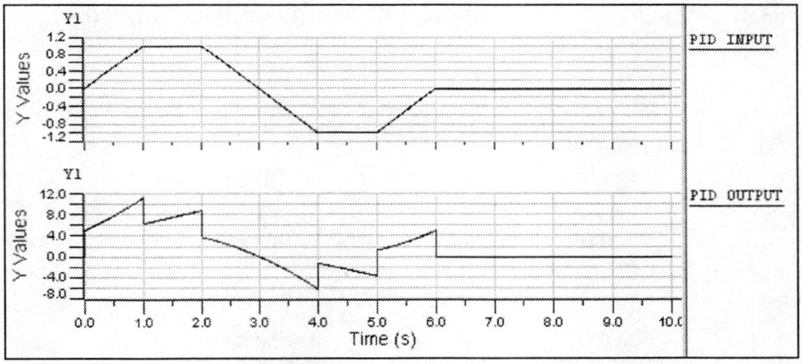

Figure 6-50. Results of PID controller test (KR=5.0, TN=2.0, TV=1.0)

6.5 Mixed-Signal Simulation

A mixed-signal simulation must always be carried out if the system being simulated includes both analog and digital parts. In VHDL-AMS analog and digital behavior (simultaneous and concurrent statements, respectively) can be described within the same architecture. You must consider:

- Method of description to allow exchanging data between the analog and the digital part of a model
- An extended simulation cycle that takes into account the solving of DAE's (*differential algebraic equations* but leaves the previous VHDL simulation cycle untouched
- Synchronization between analog and digital simulators due to different time scales.

6.5.1 Attributes for mixed-signal modeling

Interaction between analog and digital parts

The handling of analog-digital, or in other words mixed-signal systems, is one of the advantages of VHDL-AMS. Besides the introduction of standard statements to describe the analog behavior this is another innovative feature of VHDL-AMS.

Let us consider some basic ideas of how to handle mixed-signal systems in order to understand the language constructs required to describe mixed-signal systems. Analog and digital parts of a model are solved with different algorithms. The solution of the analog part can depend on the values of digital signals. You can imagine that at an analog solution time point the analog solver reads the values of digital signals. Between digital events the values of signals do not change. On the other hand digital signals may depend on analog quantities.

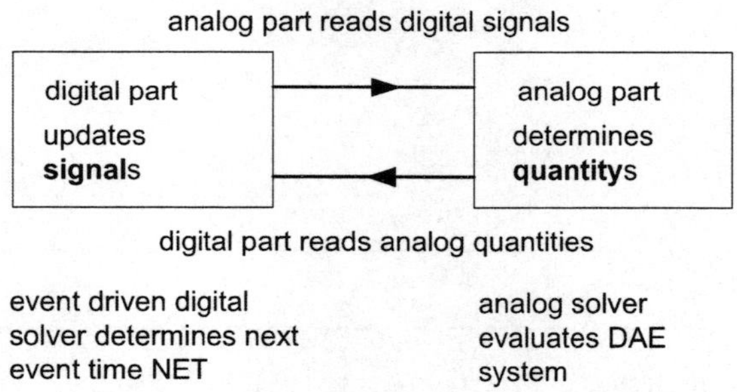

Figure 6-51. Exchange of values between analog and digital parts

How are these ideas supported by the language? The following problems have to be considered:

- If a signal is read by the analog part a signal transaction may cause a discontinuity in the analog part. To overcome this problem the following approaches are possible:

  - The analog solver must be given a hint that a discontinuity can occur. This happens for instance if a quantity directly equals a real-valued signal. The discontinuity can be announced with the **break** statement as will be shown.
  - Discontinuities can be avoided if quantities do not immediately follow signals. The changes can be carried out within a given time or with a

given slope. This is supported by 'RAMP and 'SLEW attributes that can be applied on real-valued signals.

- The digital event time points are normally only determined by the event driven solution algorithms. Only at these time points can new values of quantities be evaluated in the digital part. To force the event-driven algorithm to evaluate quantity values earlier, an event has to be generated. This can be done using 'ABOVE attribute.
- The interactions and the order of calling the analog and digital solvers have to be determined by the standard. This defines the mixed-signal simulation cycle.

Attribute 'ABOVE

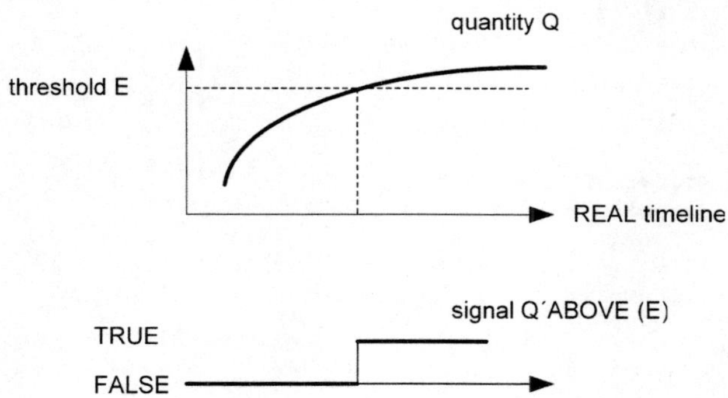

Figure 6-52. Quantity Q and BOOLEAN signal Q'ABOVE(E)

The 'ABOVE attribute is characterized in the following way:

- Q'ABOVE(E) is a BOOLEAN signal that is TRUE if the scalar quantity Q is greater than the value of E, or FALSE if Q is less than E. Otherwise, the value does not change.
- E is a real-valued expression. Any quantity appearing in this expression must be denoted by a static name. The expression is not required to be static.

Example

The following model converts an electrical voltage to a digital signal of type STD_LOGIC.

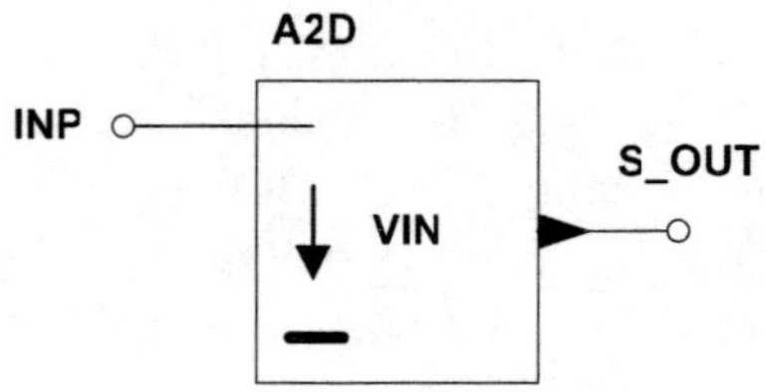

Figure 6-53. Interface of a simple A/D converter (electrical input INP)

```
library IEEE;
    use IEEE.ELECTRICAL_SYSTEMS.all;
    use IEEE.STD_LOGIC_1164.all;

entity A2D is
  generic (LEVEL : REAL := 2.5;    -- threshold level [V]
           HYST  : REAL := 0.0);   -- hysteresis [V]
  port    (terminal INP    : ELECTRICAL;
           signal   S_OUT : out STD_LOGIC := '0');
begin
  assert HYST >= 0.0
    report "ERROR: Hysteresis HYST >= 0.0 required."
    severity ERROR;
end entity A2D;

architecture IDEAL of A2D is
  quantity VIN across INP;
begin
  S_OUT <= '1' when VIN'ABOVE(LEVEL) else '0';
end architecture IDEAL;

architecture EXTENDED of A2D is
  quantity VIN across INP;
begin
  S_OUT <= '1' when VIN'ABOVE(LEVEL + HYST/2.0) else
           '0' when not VIN'ABOVE(LEVEL - HYST/2.0);
end architecture EXTENDED;
```

An open branch connects the electrical terminal INP and the electrical reference. VIN measures the input voltage. If the value of VIN is above the value of LEVEL, the signal S changes to ‘1’ in the architecture IDEAL. Otherwise the output signal value is set to ‘0’. The BOOLEAN signal VIN'ABOVE(LEVEL) is used as the condition in the conditional signal assignment statement. The architecture EXTENDED works in a similar way. However, the threshold values are determined by LEVEL +/- half of the hysteresis parameter HYST.

Attribute 'RAMP

The 'RAMP attribute is characterized in the following way:

- Suppose S is a signal of floating point type.
- S'RAMP(TRISE, TFALL) is a quantity where each scalar subelement follows S. If S changes its value, S'RAMP changes the value with rising time TRISE and falling time TFALL.
- TRISE is a static expression of floating point type that evaluates to a nonnegative number. If omitted, it defaults to 0.0. TFALL is a static expression of floating point type that evaluates to a nonnegative number. If omitted, it defaults to TRISE.
- S'RAMP is an implicit quantity. It must not be declared.

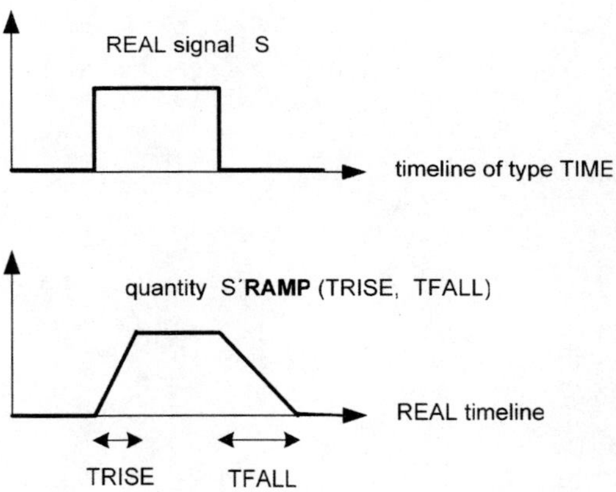

Figure 6-54. Signal S and quantity S'RAMP(TRISE, TFALL)

The signal S must be initialized otherwise numerical problems can occur in the initialization phase. If no explicit initial value is specified the default value of a signal S is S'LEFT. This would also be the initial value of S'RAMP. If S is of type REAL, the initial value would be REAL'LEFT (this is approximately -1.0E38). Analog solution points cannot normally be found for such values.

Example

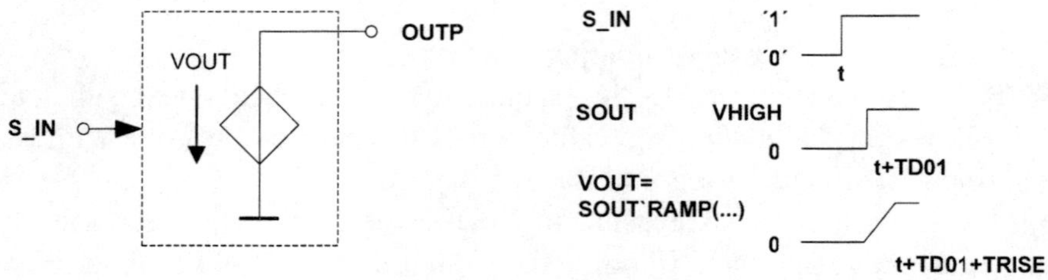

Figure 6-55. Interface and behavior of a simple D/A converter

```
library IEEE;
    use IEEE.ELECTRICAL_SYSTEMS.all;
    use IEEE.STD_LOGIC_1164.all;

entity D2A is
    generic (VHIGH : REAL := 5.0;  -- high voltage [V]
             TD_01 : TIME := 0 ns; -- posedge delay
             TD_10 : TIME := 0 ns; -- negedge delay
             TRISE : REAL := 0.0;  -- rising time [s]
             TFALL : REAL := 0.0); -- falling time [s]
    port
     (signal   S_IN  : in STD_LOGIC;
      terminal OUTP  : ELECTRICAL);
begin
    assert VHIGH > 0.0 and TRISE >= 0.0 and TFALL >= 0.0
       report "ERROR: Wrong parameters."
       severity ERROR;
end entity D2A;

architecture IDEAL of D2A is
   quantity  VOUT across IOUT through OUTP;
   signal    SOUT : REAL  := 0.0;
begin
   SOUT <= VHIGH after TD_01 when To_Bit(S_IN) ='1' else
           0.0   after TD_10;

   VOUT == SOUT'RAMP (TRISE, TFALL);
end architecture IDEAL;
```

The input signal S_IN updates the real-valued internal signal SOUT. Amplitude VHIGH and delay times TD_01 and TD_10 are taken into consideration. The output voltage source with value VOUT between the electrical terminal OUTP and the electrical reference node is derived from SOUT using the 'RAMP attribute.

Attribute 'SLEW

The 'SLEW attribute is characterized in the following way:

- Suppose S is a signal of floating point type.
- S'SLEW(RISING_SLOPE, FALLING_SLOPE) is a quantity where each scalar subelement follows S. If S changes its value, S'SLEW changes the value with rising slope RISING_SLOPE and falling slope FALLING_SLOPE.
- RISING_SLOPE is a static expression of floating point type that evaluates to a positive number. If omitted, it defaults to REAL'HIGH. FALLING_SLOPE is a static expression of floating point type that evaluates to a negative value. If omitted, it defaults to RISING_SLOPE.
- S'SLOPE is an implicit quantity. It must not be declared.

The signal S must be initialized. The reasons were discussed in the notes for the 'RAMP attribute. If no explicit initial value is specified the default value of a real-valued signal is REAL'LEFT. This would also be the initial value of S'SLEW in the first iteration step. Analog solution points cannot normally be found for such values.

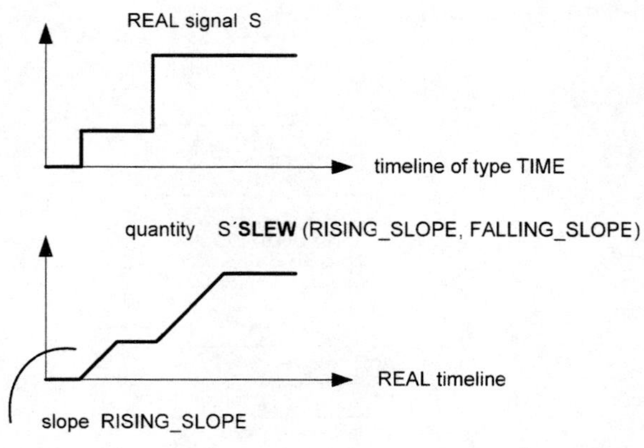

Figure 6-56. Signal S and quantity S'SLEW(RISING_SLOPE, FALLING_SLOPE)

Concurrent break statement

'RAMP and 'SLEW attributes support the digital to analog interaction. A smooth change of analog waveforms which depend on digital signals is achieved if these attributes are applied. However, there are other situations where the continuity of analog waveforms cannot be assured if digital signals change their values.

This can occur, for instance, if a real-valued S is directly used in a simultaneous statement. An example is

```
V == S;
```

Here V is a branch voltage. In this case, a discontinuity of V results from an event on S. Such a situation can produce problems for the analog solver. This is one reason the break statement is available in VHDL-AMS. It announces the possibility of discontinuities to the analog solver. The simplest form is

```
break on signal_list ;
```

That means a discontinuity is notified if an event occurs in one of the signals of the signal list. In this manner discontinuities of the derivatives of quantities can also be indicated. The break is implicitly included if 'ZOH, 'ZTF, 'RAMP, and 'SLEW attributes are used. The concurrent form can be used like a concurrent statement.

The break statement can be extended by conditions and requirements for initial conditions in the initialization phase and after discontinuities. However, this is beyond the scope of this introduction.

6.5.2 Mixed-signal simulation cycle

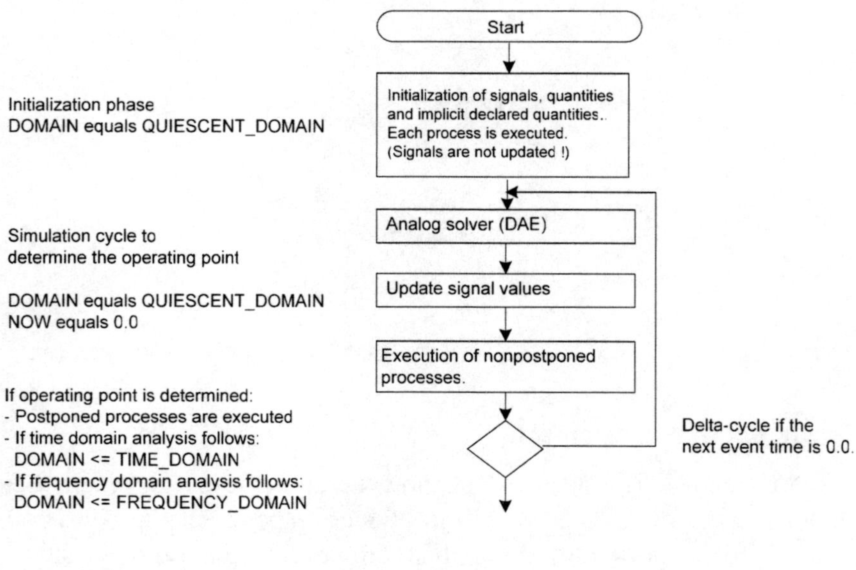

Figure 6-57. Initialization phase

After the elaboration of a design the initialization phase starts. The operating point is determined. The following rules are applied

- The DOMAIN signal is set to QUIESCENT_DOMAIN.
- The time NOW is set to 0.0.
- The initial values for the determination of signals S and quantities are by default S'LEFT and 0.0 respectively.
- At the beginning the analog solver attempts to find a solution for the given initial signal values. That is why signals that are used in simultaneous statements must be initialized explicitly.
- Afterwards the digital solver determines new values of digital signals.
- If analog and digital values are not in accordance the analog solver is called again.
- Otherwise the simulation continues with time or frequency domain simulation.

As a consequence of the simulation cycle, real signals, which are used in simultaneous statements, should be initialized. Otherwise, the analog solver has to take into consideration their default initial values REAL'LEFT during the first call. This produces numerical problems.

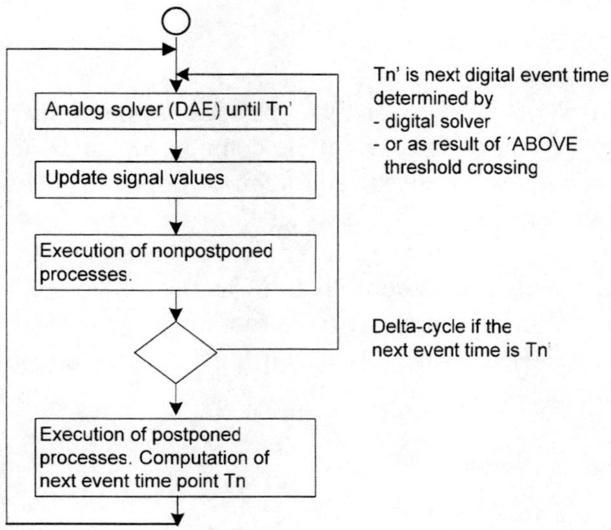

Figure 6-58. Simulation in the time domain (principle)

During the time domain simulation the DOMAIN signal is set to TIME_DOMAIN. The simulator repeats the simulation cycle shown Figure 6-58. It is done in the following way

- Tc is the current simulation time. Tn is the next time point where a digital event can be found in the digital event queue.
- The analog solver starts to simulate from Tc to Tn. Remember that between event time points the signals do not change their values. That means that the analog solver does not have to wait for new results from the digital solver.
- If Q'ABOVE(E) is used and Q crosses E then an event is placed into the event queue. Tn is set to Tn', that is the time point where the crossing occurs.
- At Tn digital signals and analog quantities are determined within a so-called delta-cycle until they are in accordance.
- After the delta-cycle is stable the analog solver can again continue until the new next event time point.

Digital time values are of type TIME. Analog time values are of type REAL. The following expressions allow converting TIME values to REAL values and vice versa:

```
REAL (TIME'POS (time_of_type_time)) * 1.0E-15   => time of type REAL
INTEGER (time_of_type_real * 1.0E15) * 1 fs     => time if type TIME
```

6.6 Analysis Domains

In order to describe and to analyze physical systems the underlying domains have to be considered. The most common domains are operating point analysis, time domain analysis, and frequency domain analysis.

Other domains are used for special analyses (for example noise simulation) or for a special kind of systems (for example periodic steady state analysis for RF circuits as described in Chapter 3 “Simulation Tools in System Design”). In the following we will describe the analysis domains that are supported by VHDL-AMS. We will pay special attention to the frequency domain analyses.

6.6.1 Supported domains

VHDL-AMS supports different kinds of analyses. The DOMAIN signal is updated by the simulation engine, which provides a hint of which kind of analysis is used. The following kinds of analyses are possible

- *Determination of operating point*
  Quiescent domain analysis is used to determine the operating point. The DOMAIN signal is set to QUIESCENT_DOMAIN. The time NOW

equals zero (0.0 or 0 fs). By default, the set of characteristic expressions is augmented by

- Q'DOT = 0.0
  if Q'DOT is used somewhere in the simultaneous statements.
- Q = 0.0
  if Q'INTEG is used somewhere in the simultaneous statements.
- Q'DELAYED(T) – Q = 0
  if Q'DELAYED(T) is used somewhere in the simultaneous statements.

The operating points determine the initial values for time domain simulation.

- *Time domain analysis*
  Time domain analysis is used to determine the behavior of a system over time. It begins after the determination of the operating point at time NOW=0.0. The DOMAIN signal is set to TIME_DOMAIN.
- *Frequency domain analysis*
  Frequency domain analysis is used to determine the small signal behavior for sinusoidal waveforms around the operating point. After quiescent domain analysis the network equations are linearized at the operating point. The characteristic expressions are derived by the simulation engine from the time domain simultaneous statements using linearization. Thus, frequency domain analysis can be applied for linear systems or systems that are linearized at the operating point, and systems with sinusoidal sources where all sources are evaluated at the same frequency. The results of the frequency domain analysis only describe the steady state. Frequency analysis can be done as small signal frequency domain analysis or small signal noise analysis.

In conclusion, in the execution phase of a VHDL-AMS simulation, a quiescent domain analysis is performed first. Immediately after quiescent domain analysis either time domain analysis or frequency domain analysis is performed. Time domain analysis uses the result of the quiescent domain analysis as its initial value. Frequency domain analysis uses a system that is linearized at the operating point. Table 6-5 shows how the kinds of analysis in VHDL-AMS simulation engines are in close relation with those found in SPICE-like simulation engines.

Table 6-5. Kinds of analysis

Kinds of analysis	VHDL-AMS simulation engine	SPICE-like simulation engine (cicuit level simulator)
Determination of operating point	DOMAIN equals QUIESCENT_DOMAIN	Determination of DC (direct current) operating point (.OP)
Time domain analysis	DOMAIN equals TIME_DOMAIN	Transient analysis (.TRAN)
Small signal frequency domain analysis	DOMAIN equals FREQUENCY_DOMAIN; Sinusoidal spectrum sources are described by magnitude and phase	AC (alternating current) analysis (.AC)
Small signal noise domain analysis	DOMAIN equals FREQUENCY_DOMAIN; Noise sources are described by their spectral density	Noise analysis (.NOISE)

6.6.2 Small-signal and noise domain simulation

AC analysis basics

Linear systems that are described by linear differential systems of equations can be handled in a special way. In AC analysis (*alternating current*), we assume that independent voltage and current sources (across and through respectively) are sinusoidal waveforms of the same frequency. In the steady state, all quantities are also sinusoidal waveforms of the same frequency. The notion of AC analysis is to solve a linear system of complex equations instead of the linear system of differential equations.

Table 6-6. Correspondence between time and frequency domains

	Time domain	Small signal analysis (AC)
General waveform description	x	X
Sinusoidal waveform with known frequency $\omega=2\pi f$	$C\cdot\cos(\omega\cdot t+\varphi)$	$C\cdot e^{j\,\varphi}$
Differentiation	$x\cdot d/dt$	$j\,\omega\cdot X$
Integration	$\int x$	$X\cdot 1/j\,\omega$
Delay for waveform x of known frequency $\omega=2\pi f$	$x(t-T)$	$X\cdot e^{-j\,\omega T}$

In the AC calculus a complex linear system of equations is assigned to the system of linear differential equations applying the rules given in the table. This complex system of equations is solved. The results can be interpreted as time domain waveforms. The transformation and solving of the systems of equations is done for a special (fixed) frequency. Figure 6-59 illustrates this approach.

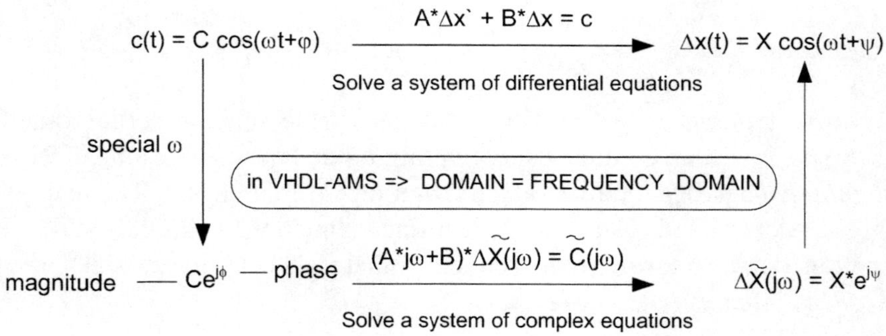

Figure 6-59. System with sinusoidal input

What are the consequences of taking into consideration the modeling requirements? After determination of the operating point the linearization of the network equations can be done automatically by the simulation engine (see Table 6-6). That means that in general no special description for frequency analysis is necessary. The description for frequency analysis can be derived from the description for time domain analysis. Only AC voltage and current sources (across and through respectively) that are characterized by magnitude and phase have to be added to the description for frequency domain analysis.

The AC linear circuit is analyzed over a user-specified range of frequencies. The frequencies normally start with frequency *fstart* and step with a specified number of points per decade to the final frequency *fstop* (DEC mode). Alternatively, they step with a specified number of points per octave to the final frequency *fstop* (OCT mode) or step with a specified number of points that are linearly distributed between *fstart* and *fstop*.

This AC method was introduced into electrical engineering by Charles Proteus Steinmetz at the end of the 19th century. For the first time it offered a widely accepted way to design alternating current electrical equipment using mathematical methods. The frequency domain representation of a sinusoid is usually called *phasor*.

Declaration of a spectral source quantity

Spectral source quantities can be declared in an architecture body with a spectral source quantity declaration. The general form of the declaration is

```
quantity q_name : real_type spectrum magnitude, phase;
```

where *q_name* is an identifier that names the spectral source quantity. *real_type* is scalar or composite floating-point type (for example REAL). *magnitude* specifies the real valued magnitude of the phasor. The magnitude can be determined using the predefined function FREQUENCY. This function delivers the current frequency that is used for spectral analysis. *phase* specifies the phase of *q_name*.

The quantity *q_name* can be applied in simultaneous statements. It is only taken into consideration if the DOMAIN signal equals FREQUENCY_DOMAIN otherwise it is set to zero.

Example

To determine the transfer characteristic of a circuit we use an input voltage with magnitude one and phase zero. This voltage source is connected to UUT1 (the linear analog lowpass filter from Section 6.3.5) and UUT2 (the linear digital lowpass filter from Section 6.3.5). You can observe the transfer characteristic of UUT1 at node N_ANA_OUT and the transfer characteristic of UUT2 at node N_DIG_OUT (see Figure 6-60).

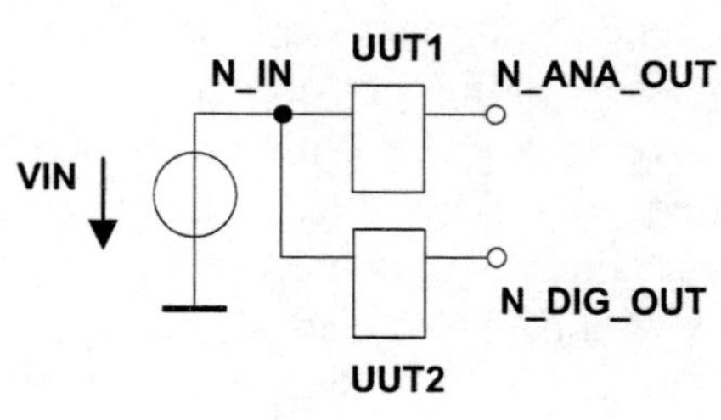

Figure 6-60. Test circuit for AC analysis

In the VHDL-AMS description of the test circuit, a branch with branch quantity VIN is declared that connects terminal N_IN and the electrical reference node. A spectrum quantity V_AC is declared with magnitude 1 and phase 0. For frequency domain simulation VIN equals V_AC. Otherwise it is described by a sinusoidal voltage waveform in the time domain. Other waveforms are of course also possible in the time domain. The cut-off frequency of both filters is 2 kHz, and the sampling frequency of the digital filter is 8 kHz. Thus, the first break in the magnitude of the digital filter is at 4 kHz.

```
library IEEE;
    use IEEE.ELECTRICAL_SYSTEMS.all;  use IEEE.MATH_REAL.all;
    use WORK.all;    -- for UUT1 and UUT2

entity BENCH is end entity BENCH;

architecture BENCH_AC of BENCH is
    terminal N_IN       : ELECTRICAL;
    terminal N_ANA_OUT : ELECTRICAL;      -- output analog filter
    terminal N_DIG_OUT : ELECTRICAL;      -- output digital filter
    quantity VIN  across IIN through N_IN;
    quantity V_AC : REAL spectrum 1.0, 0.0;    -- spectrum source
begin
    if DOMAIN /= FREQUENCY_DOMAIN use
       VIN == SIN(MATH_2_PI*1.0E3*NOW);        -- other waveforms
    else
       VIN == V_AC;
    end use;

UUT1: entity LOWPASS(BESSEL_2)
       generic map (FC => 2.0E3)
       port map (INP => N_IN, OUTP => N_ANA_OUT);
UUT2: entity LOWPASS(BESSEL_2_DIGITAL)
       generic map(FC => 2.0E3)
       port map (INP => N_IN, OUTP => N_DIG_OUT);
end architecture BENCH_AC;
```

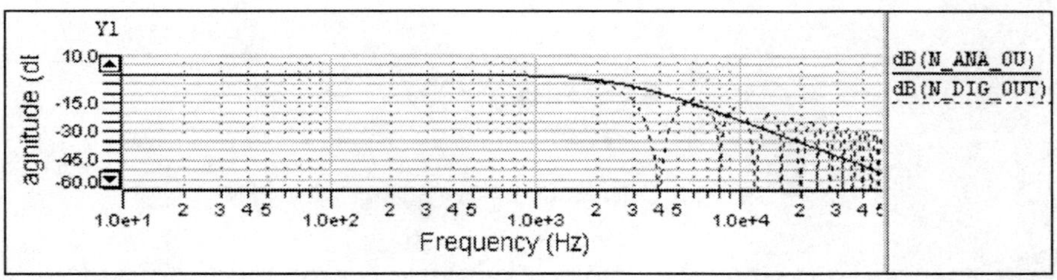

Figure 6-61. Frequency response of analog and digital filter

Noise analysis basics

Small signal noise analysis presumes:

- Small signal noise analysis can be applied to linear circuits and circuits that are linearized around an operating point.
- Amplitudes of noise sources are Gaussian distributed. The variances of the distributions are constant over time. The mean values are supposed to be zero.

- Noise sources are characterized by their noise spectral density. All noise spectral densities S_v and S_i (voltages and currents respectively) are in squared units (V^2/Hz and A^2/Hz for spectral density).
- In the case of a noise source the relation between spectral noise density $S(f)$ and the root mean square (effective value) V_{eff} is given by

$$V_{eff}^2 = \int_{f_{start}}^{f_{stop}} S(f)df$$

f_{start} and f_{stop} are the limits of the frequency range that has to be taken into consideration to describe the noise source. In the case of constant spectral density S it follows

$$V_{eff}^2 = S \cdot (f_{stop} - f_{start}) = S \cdot \Delta f$$

- The spectral density S_A of a random waveform that results from a source with spectral density S_E is given by the squared transfer function times S_E, see Figure 6-62 ($\omega = 2\pi f$).

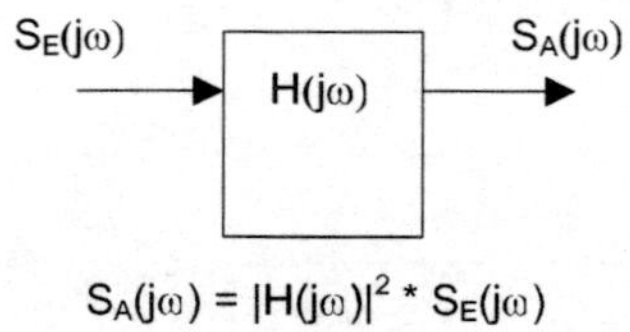

Figure 6-62. Calculation of spectral noise densities

- The spectral densities of uncorrelated noise contributions are added (see also [Std99], Section 12.8).

In a simulation engine noise analysis is done in a specified frequency range.

Declaration of a noise quantity source

Noise quantities are characterized in the following way: Noise quantities can be declared in an architecture body with a noise quantity declaration. The general form of the declaration is

```
quantity q_name : real_type noise density;
```

where *q_name* is an identifier that names the quantity. *real_type* is scalar or composite floating-point type (for example REAL). *density* specifies the spectral noise density depending on the predefined function frequency. If it does not depend on the frequency, the spectral density is constant over the frequency range.

The quantity *q_name* can be applied in simultaneous statements. It is only taken into consideration if the DOMAIN signal equals FREQUENCY_DOMAIN otherwise it is set to zero.

Example

Thermal noise of a resistor can be modeled as shown in Figure 6-63. A current noise source is in parallel with the noiseless resistor (see for example [LuB00], Appendix H). The noise source is characterized by a constant spectral density $S_I = \dfrac{4kT}{R}$. k is the Boltzmann constant ($1.38 \cdot 10^{-23} \dfrac{Ws}{K}$). T is the absolute temperature in degree Kelvin and R the value of the resistor. In the VHDL-AMS model Boltzmann's constant equals PHYS_K which is declared in the IEEE library in package FUNDAMENTAL_CONSTANTS.

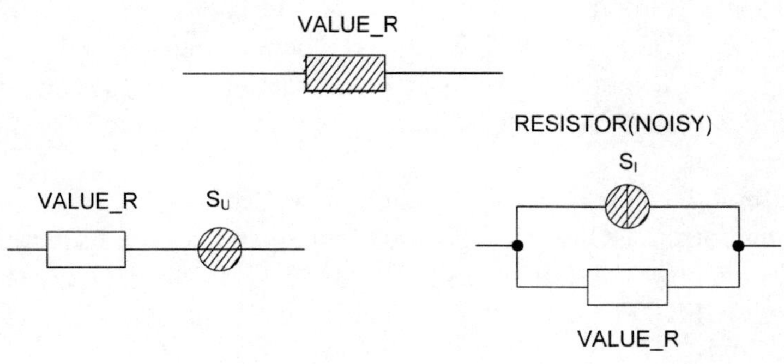

Figure 6-63. Noise models of a linear resistor

```
library IEEE;
   use IEEE.ELECTRICAL_SYSTEMS.all, IEEE.FUNDAMENTAL_CONSTANTS.all;

entity RESISTOR is
   generic (VALUE_R         : REAL := 1.0;     -- resistance [Ohm]
            TEMP            : REAL := 300.0); -- temperature [K]
   port    (terminal P1, P2 : ELECTRICAL);
begin
   assert VALUE_R > 0.0 and TEMP > 0.0
      report "ERROR: Parameters are not correct." severity ERROR;
end entity RESISTOR;

architecture NOISY of RESISTOR is
    quantity V  across I, INOISE through P1 to P2; -- parallel br.
    quantity SI : REAL noise 4.0*PHYS_K*TEMP/VALUE_R;
begin
    V      == VALUE_R * I;
    INOISE == SI;
end architecture NOISY;
```

6.7 Summary

In this chapter we gave a short introduction to VHDL-AMS. The basic language constructs and ideas of VHDL-AMS were explained. More information can be found in the IEEE Std 1076.1. We focused on the analog extension of VHDL-AMS compared to the pure digital oriented VHDL. Some new features in VHDL-AMS consist of:

- Language constructs to describe conservative network semantics (terminals, natures, branch quantity declarations)
- Language constructs to describe nonconservative signal flow semantics (quantity ports)
- Simultaneous statements to describe analog constitutive relations
- Special support to model analog behavior ('DOT, 'INTEG, 'DELAYED, 'LTF, 'ZOH, 'ZTF, and other attributes)
- Support to model mixed-signal interaction between analog and digital parts ('ABOVE, 'RAMP, 'SLEW attributes, and break statement)
- Definition of the mixed-signal simulation cycle
- Usage of the DOMAIN signal to distinguish between different kinds of analysis (QUIESCENT_DOMAIN, TIME_DOMAIN, and FREQUENCY_DOMAIN including small signal AC and noise analysis)

In this chapter we did not consider facilities to model multi-domain systems consisting of electrical and nonelectrical parts. Some other language extensions were also not considered (for example the full break statement).

To readers interested in these features we recommend the VHDL-AMS standard [Std99] and other textbooks (for example [APT03]) for further reading.

Chapter 7

SELECTED RF BLOCKS IN VHDL-AMS

7.1 Library Overview

The previous chapter introduced some fundamental concepts and the description syntax of the VHDL-AMS language. This chapter presents a library of typical RF building blocks in VHDL-AMS.

The models are subdivided into three categories of blocks:

- Signal sources
- System blocks for signal processing
- Measurement and observation units

For all blocks behavioral models with RF specific properties are provided. They are uniformly documented with:

- Functional description
- Model interface
- Model implementation
- Simulation example with results if necessary

In Chapter 9 a complex model of a WLAN receiver is assembled from these basic building blocks. The chapter describes how to instantiate and parameterize the models, and how to run the simulation.

7.2 Signal Sources

Table 7-1. Signal source blocks overview

Model	Properties
Independent voltage sources	Sinusoidal source as single- or two-tone source
Modulated sources	AM or FM sinusoidal source
Wobble generator	Sinusoidal source with swept frequency
Pseudorandom binary source	Feedback shift register with variable length

7.2.1 Independent sources

Functional description

In this section different sources are provided in a SPICE-like notation but with RF-specific extensions. They are named *p_sin* since the sinusoidal output of the source is specified in terms of power. Also, in contrast to SPICE sources, a positive and finite value for the output resistor is required. By calling different architectures it is possible to decide whether to use a single- or a two-tone source. Refer to the next section for further architectures of the same source.

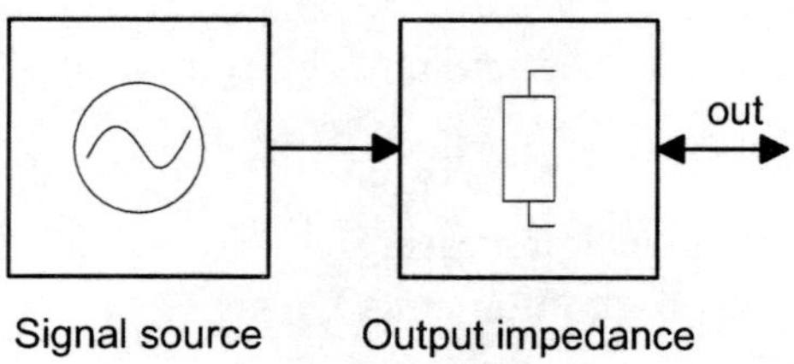

Figure 7-1. Block diagram of an RF-specific source model

The main characteristics of the independent source models are as follows.

- A sinusoidal source is modeled with single-tone output

$$v_{src} = 2 \cdot [vo + va \cdot \sin(2\pi \cdot freq \cdot t + phase)]$$

where *va* denotes the voltage amplitude that is computed from the power amplitude parameter *pa_dBm* by

$$va = \sqrt{2 \cdot rout \cdot 10^{\frac{pa\_dBm - 30}{10}}}$$

- A sinusoidal source with two-tone output is also available

$$v_{src} = 2\cdot[vo + va\cdot\sin(2\pi\cdot freq\cdot t + phase) + va2\cdot\sin(2\pi\cdot freq2\cdot t + phase2)]$$

where *va* and *va2* denote the voltage amplitudes that are computed from the power amplitude parameters *pa_dBm* and *pa2_dBm* respectively by

$$va = \sqrt{2\cdot rout\cdot 10^{\frac{pa\_dBm-30}{10}}}\ ,\ va2 = \sqrt{2\cdot rout\cdot 10^{\frac{pa2\_dBm-30}{10}}}$$

- Note: A factor of two is added in both cases since a power source has its maximum power output when used in a matched system, where an external resistor equal to the internal resistance is connected. In this case the specified power amplitude can be measured.
- The impedance of the output port is modeled as an ohmic resistance. A typical value of a matched system is:

$$R_{out} = 50\Omega$$

Model interface

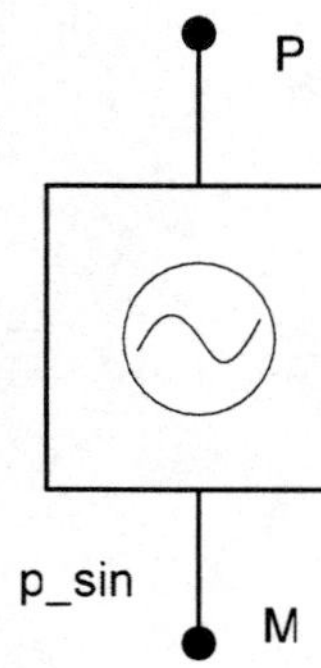

Figure 7-2. Schematic symbol of the source model

Table 7-2. Model ports

Name	Type	Description
P	ELECTRICAL	Positive pin
M	ELECTRICAL	Negative Pin

Table 7-3. Model parameters

Name	Unit	Default value	Description
VO	V	0.0	Offset voltage
PA_DBM	dBm	-100.0	Power amplitude of sine wave
FREQ	Hz	1.0e03	Frequency of sine wave
PHASE	rad	0.0	Phase of sine wave
PA2_DBM	dBm	-100.0	Power amplitude of sine wave (second tone)
FREQ2	Hz	1.0e03	Frequency of sine wave (second tone)
PHASE2	rad	0.0	Phase of sine wave (second tone)
ROUT	Ohm	50.0	Output resistance

Model implementation

```
architecture SINGLE_TONE of P_SIN is
  constant PA:     REAL:= 10**((PA_DBM-30.0)/10.0);
  constant PA:     REAL:= SQRT(PA * 2.0 * ROUT);
  terminal N_INT: ELECTRICAL;
  quantity V_ROUT across I_ROUT through P to N_INT;
  quantity V_SRC  across I_SRC  through N_INT to M;
begin

  -- signal source
  V_SRC == 2.0 * (VO + VA * SIN(NOW * MATH_2_PI*FREQ + PHASE));

  -- output port resistance
  V_ROUT == ROUT * I_ROUT;

end architecture SINGLE_TONE;
```

The complete model is included on the CD-ROM that is provided with this book.

Simulation example

For a simulation example using independent sources see Section 7.3.2.

7.2.2 Modulated sources

Functional description

In this section different sources are provided in a SPICE-like notation but with RF-specific extensions. They are named *p_sin* since the sinusoidal output of the source is specified in terms of power. Also, in contrast to SPICE sources, a positive and finite value for the output resistor is required. By calling different architectures it is possible to decide whether to use an amplitude modulated (AM) or a frequency modulated (FM) sinusoidal source. Refer to the previous section for further architectures of the same source.

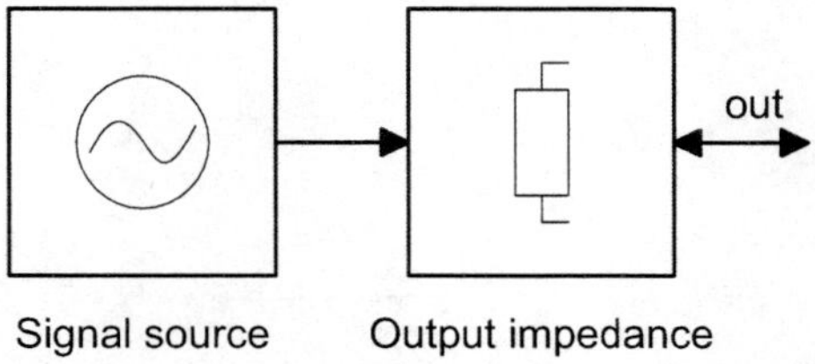

Figure 7-3. Block diagram of an RF-specific source model

The main characteristics of the modulated source models are as follows.

- A sinusoidal source is modeled with amplitude modulation (AM)

$$v_{src} = 2 \cdot \left[vo + va \cdot (1 + mdi \cdot \sin(2\pi \cdot freqm \cdot t + phasem)) \sin(2\pi \cdot freq \cdot t + phase) \right]$$

where *va* denotes the voltage amplitude that is computed from the power amplitude parameter *pa_dBm* by

$$va = \sqrt{2 \cdot rout \cdot 10^{\frac{pa\_dBm - 30}{10}}}$$

- Another architecture for the sinusoidal source is included with frequency modulation (single frequency FM - SFFM)

$$v_{src} = 2 \cdot \left[vo + va \cdot \sin\left(2\pi \cdot freq \cdot t + phase + mdi \cdot \sin(2\pi \cdot freqm \cdot t + phasem)\right) \right]$$

where *va* denotes the voltage amplitude that is computed from the power amplitude parameter *pa_dBm* by

$$va = \sqrt{2 \cdot rout \cdot 10^{\frac{pa\_dBm - 30}{10}}}$$

- Note: A factor of two is added in both cases since a power source has its maximum power output when used in a matched system, where an external resistor equal to the internal resistance is connected. In this case the specified power amplitude can be measured.
- The impedance of the output port is modeled as an ohmic resistance. A typical value of a matched system is:

$$R_{out} = 50\Omega$$

Model interface

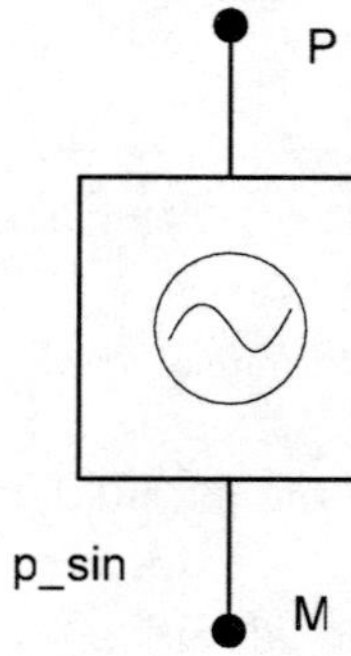

Figure 7-4. Schematic symbol of the source model

Table 7-4. Model ports

Name	Type	Description
P	ELECTRICAL	Positive pin
M	ELECTRICAL	Negative Pin

Table 7-5. Model parameters

Name	Unit	Default value	Description
VO	V	0.0	Offset voltage
PA_DBM	dBm	-100.0	Power amplitude of sine wave
FREQ	Hz	1.0e03	Frequency of sine wave
PHASE	rad	0.0	Phase of sine wave
MDI	-	0.0	Modulation index
FREQM	Hz	1.0	Modulation frequency
PHASEM	rad	0.0	Modulation phase
ROUT	Ohm	50.0	Output resistance

Model implementation

```
architecture SFFM of P_SIN is
  constant PA:     REAL:= 10**((PA_DBM-30.0)/10.0);
  constant VA:     REAL:= SQRT(PA * 2.0 * ROUT);
  terminal N_INT: ELECTRICAL;
  quantity V_ROUT across I_ROUT through P to N_INT;
  quantity V_SRC  across I_SRC  through N_INT to M;
begin

  -- signal source
  V_SRC == 2.0 * (VO + VA * SIN(NOW * MATH_2_PI*FREQ + PHASE
                  + MDI*SIN(NOW * MATH_2_PI*FREQM + PHASEM)));
  -- output port resistance
  V_ROUT == ROUT * I_ROUT;

end architecture SFFM;
```

The complete model is included on the CD-ROM that is provided with this book.

Simulation example

For a simulation example using a modulated source see Section 7.4.3.

7.2.3 Wobble generator

Functional description

A sinusoidal source is provided, where the frequency of the output signal is swept over a parameter specified range.

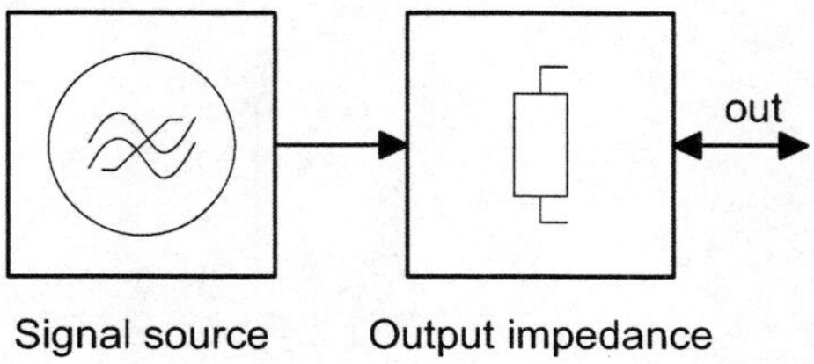

Figure 7-5. Block diagram of a wobble source model

The main characteristics of the wobble generator model are as follows.

- The sinusoidal source has a single-tone output

$$v_{src} = 2 \cdot \boldsymbol{amp} \cdot \sin(\varphi), \quad \dot{\varphi} = 2\pi f_{eff}$$

where *amp* denotes the voltage amplitude that is computed from the power amplitude parameter *amp_dBm* by

$$amp = \sqrt{2 \cdot \boldsymbol{rout} \cdot 10^{\frac{amp\_\boldsymbol{dBm}-30}{10}}}$$

and f_{eff} denotes the actual frequency

$$f_{eff} = \begin{cases} f_{start} & \text{if } t \le t_{init} \\ f_{start} + \Delta f_{sweep}(t - t_{init}) & \text{if } t > t_{init} \text{ and } f_{eff} < f_{stop} \\ f_{stop} & \text{if } f_{eff} \ge f_{stop} \end{cases}$$

- The start time of the sine wave is delayed by t_{init} = InitDelay
- The frequency changes with Δf_{sweep} = SweepRate (Hz/s)
- The Impedance of the output port is modeled as an ohmic resistance. A typical value of a matched system is:

$$R_{out} = 50\Omega$$

Model interface

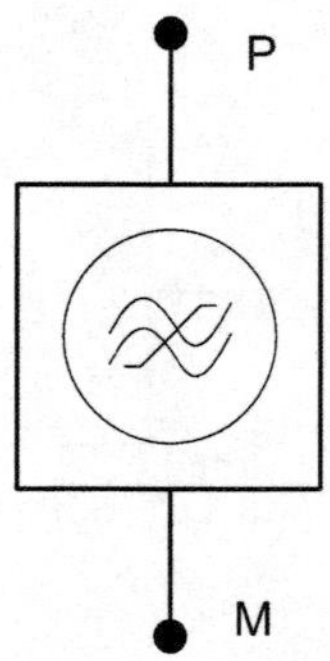

Figure 7-6. Schematic symbol for wobble generator

Table 7-6. Model ports

Name	Type	Description
P	ELECTRICAL	Positive pin
M	ELECTRICAL	Negative Pin

Table 7-7. Model parameters

Name	Unit	Default value	Description
AMP_DBM	dBm	-100.0	Power amplitude of sine wave
INITDELAY	s	0.0	Initial time delay before oscillator starts
STARTFREQ	Hz	1.0	Initial frequency where sweep starts
STOPFREQ	Hz	1.0e07	End frequency where sweep stops
SWEEPRATE	Hz/s	1.0	Rate of change for frequency sweep
ROUT	Ohm	50.0	Output resistance

Model implementation

```
architecture BHV of WOBBEL is
  constant AMP_LIN  : REAL:= 10**((AMP_DBM-30.0)/10.0);
  constant AMP      : REAL:= SQRT(AMP_LIN * 2.0 * ROUT);
  terminal N_INT    : ELECTRICAL;
  quantity PHI      : REAL;
  quantity EFFFREQ  : REAL := STARTFREQ;
```

```
    quantity V_ROUT across I_ROUT through P to N_INT;
    quantity V_SRC  across I_SRC  through N_INT to M;
  begin
     if NOW > INITDELAY and EFFFREQ < STOPFREQ use
      EFFFREQ == STARTFREQ + SWEEPRATE*(NOW-INITDELAY);
    elsif EFFFREQ >= STOPFREQ use
      EFFFREQ == STOPFREQ;
    else
      EFFFREQ == STARTFREQ;
    end use;

    if DOMAIN = QUIESCENT_DOMAIN USE
      PHI == 0.0;
    else
      PHI'DOT == MATH_2_PI*EFFFREQ;
    end use;

  -- signal source
    V_SRC == 2.0 * AMP * SIN(PHI);

  -- output port resistance
    V_ROUT == ROUT * I_ROUT;
  end architecture BHV;
```

The model implementation is included on the CD-ROM that is provided with this book.

7.2.4 Pseudorandom binary source

Functional description

This model provides a pseudorandom sequence at its binary output. It is constructed as a maximum-length feedback shift register with variable register length.

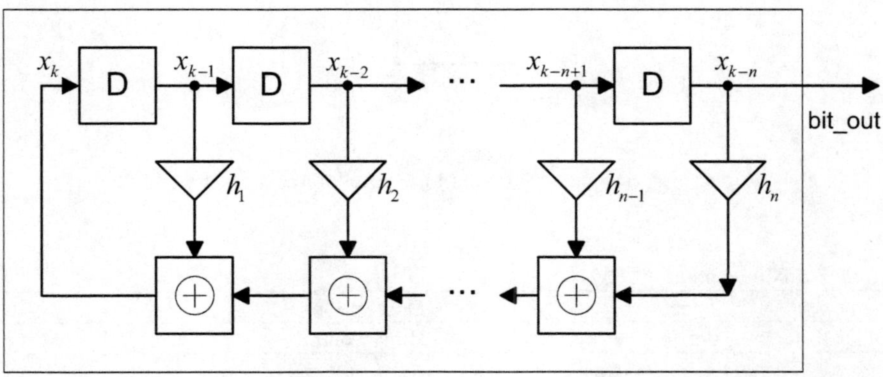

Figure 7-7. Block diagram of a pseudorandom binary source (PRBS)

The main characteristics of the modulated source models are as follows.

- The feedback shift register has a length *polygrad*, where $2 \leq polygrad \leq 34$. The generated binary sequence is of maximum length, that is it has a period of $2^{polygrad} - 1$.
- The generator polynomial for order 2 to 34 is built into the model

$$polygrad = 2: \quad x_k = x_{k-1} \oplus x_{k-2}$$
$$polygrad = 3: \quad x_k = x_{k-1} \oplus x_{k-3}$$
$$polygrad = 4: \quad x_k = x_{k-1} \oplus x_{k-4}$$
$$\vdots$$

where the addition is modulo 2. For a complete list of generator polynomials see [BLM04].

- The initial state of the shift register can be set with parameter *seed*.
- The frequency of the binary sequence is adjustable with the parameter *bit_time*.
- An initial time delay (before the sequence starts) can be applied using the parameter *bit_del*.

Model interface

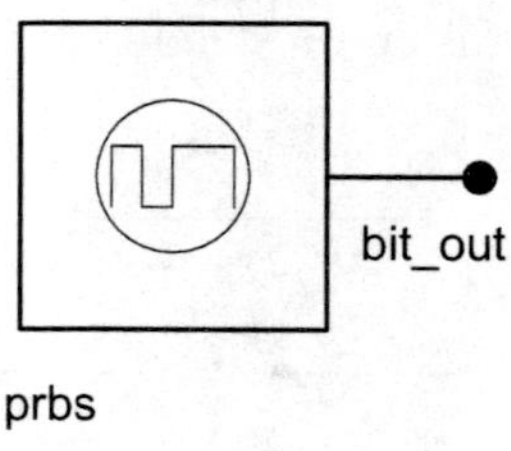

Figure 7-8. Schematic symbol of the pseudorandom binary source

Table 7-8. Model ports

Name	Type	Description
BIT_OUT	out BIT	Binary output

Table 7-9. Model parameters

Name	Unit	Default value	Description
POLYGRAD	-	2	Order of generator polynomial
SEED	-	"01"	Initial bit vector of feedback shift register
BIT_TIME	s	1us	Time duration of binary values
BIT_DEL	s	0us	Initial time delay of binary sequence

Model implementation

The model implementation is included on the CD-ROM that is provided with this book.

Simulation example

For a simulation example using a pseudorandom binary source see Section 7.3.8.

7.3 Basic RF Building Blocks

Table 7-10. Overview on basic RF building blocks

Model	Properties
Low-noise Amplifier	RF specific operational amplifier
Mixer	Gilbert mixer with LNA
Charge pump	Mixed-signal charge pump for PLL
Analog VCO	Sine wave with tunable frequency
Digital VCO	Square wave with tunable frequency
Filters	Lowpass and highpass Butterworth filters
Switch	Varying resistance with digital control
General n-bit A/D and D/A converter	Mixed-signal model with n-bit vector interface
Simple channel model	AWGN channel with time delay

7.3.1 Low-noise amplifier

Functional description

Low-noise amplifiers (LNA) are central elements in RF applications for amplifying signals over a wide frequency range with low signal to noise distortion. In contrast to low-frequency amplifiers the main focus is on the large signal behavior and the nonlinear distortion due to harmonics and intermodulation. Also power transmission and therefore matching impedances have to be considered. Figure 7-9 shows a typical block diagram of an LNA model, which is divided into input, output and transmission blocks.

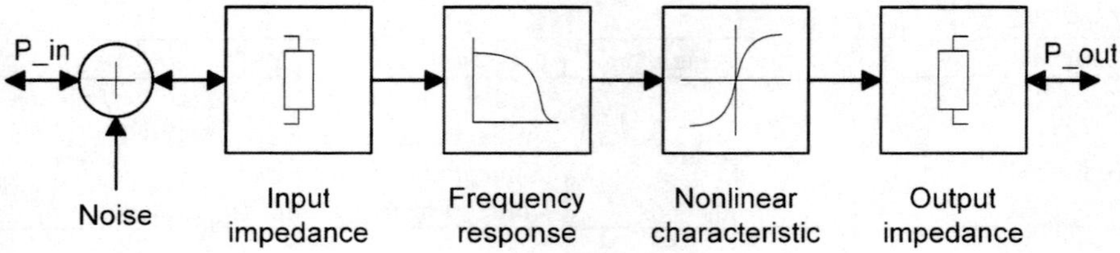

Figure 7-9. Block diagram of a simple low-noise amplifier model

The main characteristics of the low-noise amplifier model are as follows.

- Signal amplification is specified as power gain and converted to voltage gain

$$gain_{voltage,linear} = \sqrt{10^{\frac{gp\_dB}{10}} \cdot \frac{R_{out}}{R_{in}}}$$

- Nonlinear gain characteristic is expressed in terms of a 3rd order intercept point

$$v_{out} = a \cdot v_{in} - b \cdot v_{in}^3$$

where

$$a = gain_{voltage,linear},\ b = \frac{4}{3}\frac{a}{ip3^2},$$

$$ip3 = \sqrt{10^{\frac{ip3\_dBm-30}{10}} \cdot 2 \cdot R_{in}}$$

Note: The intercept point in this context refers to a single-tone signal, while for other circuits, IP2 and IP3 are measured by two tones.

- Frequency response is provided by the dominant pole (and further poles and zeros)

$$\underline{H}(j\omega) = \frac{1}{1 + j\omega / \omega_g}$$

where $\omega_g = 2\pi f_g$, and f_g is the frequency of the dominant pole.

- Input and output impedances are modeled as ohmic resistances. Typical values of a matched system are

$$R_{in} = R_{out} = 50\Omega$$

Often you will find these characteristics expressed in terms of the S-parameters.

Additional characteristics, that were not modeled here, include second order effects such as temperature and power supply dependency of the gain function, recovery time after output limitation, higher order nonlinearities, and power consumption. Also, noise contribution of the amplifier stage is not included in the model since noise quantities are not supported by ADVance MS.

Model interface

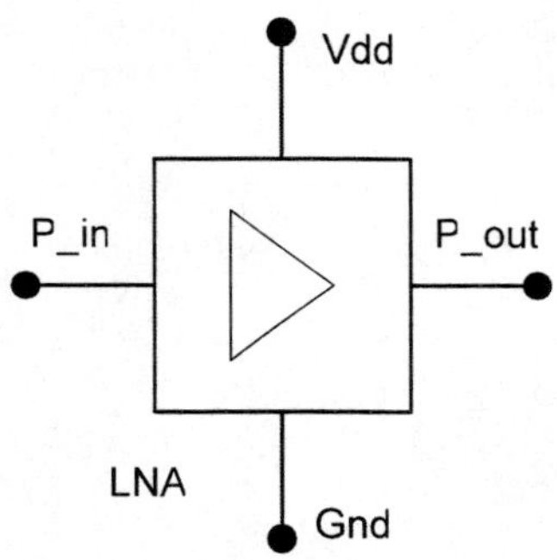

Figure 7-10. Schematic symbol of the low-noise amplifier

Table 7-11. Model ports

Name	Type	Description
P_IN	ELECTRICAL	Input pin
P_OUT	ELECTRICAL	Output pin
VDD	ELECTRICAL	Supply voltage
GND	ELECTRICAL	Reference node

Table 7-12. Model parameters

Name	Unit	Default value	Description
GP_DB	dB	0.0	Open loop power gain
IP3_DBM	dBm	-30.0	Referenced IP3
FNOISE_DB	dB	0.0	Noise figure of stage
FG	Hz	real'HIGH	Frequency of dominant pole
RIN	Ohm	50.0	Input resistance
ROUT	Ohm	50.0	Output resistance

Model implementation

```
architecture RF of LNA is
  constant GP_LIN    : REAL:= 10**(GP_DB/10.0);
                               -- linear value of power gain
  constant IP3_LIN   : REAL:= 10**((IP3_DBM-30.0)/10.0);
                               -- linear value of ip3
  constant A         : REAL:= SQRT(GP_LIN*ROUT/RIN);
                               -- linear value of voltage gain
  constant IP3       : REAL:= SQRT(IP3_LIN*2.0*RIN);
                               -- linear value of ip3 voltage
  constant B         : REAL:= A/(IP3*IP3)*4.0/3.0;
                               -- third order coefficent
  constant INMAX     : REAL:= SQRT(A/(3.0*B));
                               -- maximum input voltage for clipping
  constant OUTMAX    : REAL:= 2.0*A/3.0*INMAX;
                               -- output voltage at clipping
  terminal U_IN      : ELECTRICAL;
  terminal U_IN_F    : ELECTRICAL;
  terminal U_OUT     : ELECTRICAL;
  quantity V_NOISE across I_NOISE through P_IN to U_IN;
  quantity V_RIN   across I_RIN   through U_IN to GND;
  quantity V_UINF  across I_UINF  through U_IN_F to GND;
  quantity V_LIM   across I_LIM   through U_OUT to GND;
  quantity V_ROUT  across I_ROUT  through U_OUT to P_OUT;

begin

  -- input stage: noise figure, input resistance
  V_NOISE == 0.0;
  V_RIN == RIN * I_RIN;

  -- transmission stage:
  -- gain, voltage limitation, transfer function
  V_UINF == V_RIN'LTF((0 => 1.0), (1.0, 1.0/MATH_2_PI/FG));

  if abs(V_UINF)<INMAX use
    V_LIM == 2.0*(A - B*V_UINF*V_UINF)*V_UINF;
  elsif V_UINF > 0.0 use
    V_LIM == 2.0*OUTMAX;
  else
    V_LIM == -2.0*OUTMAX;
  end use;

  -- output stage:  output resistance
  V_ROUT == ROUT * I_ROUT;

end architecture RF;
```

The complete model is included on the CD-ROM that is provided with this book.

Simulation example

As a stimulation signal a single-tone of 1.0 kHz and –30.0 dBm was used. The LNA model was instantiated with the following parameters:

```
LNA1: entity LNA(RF)
 generic map (GP_DB => 6.0,
              IP3_DBM => -23.0,
              FNOISE_DB => 2.5
              )
 port map (P_IN => N_1,
           P_OUT => N_2,
           VDD => N_VDD,
           GND => ELECTRICAL_REF
           );
```

The complete test bench is included on the CD-ROM that is provided with this book.

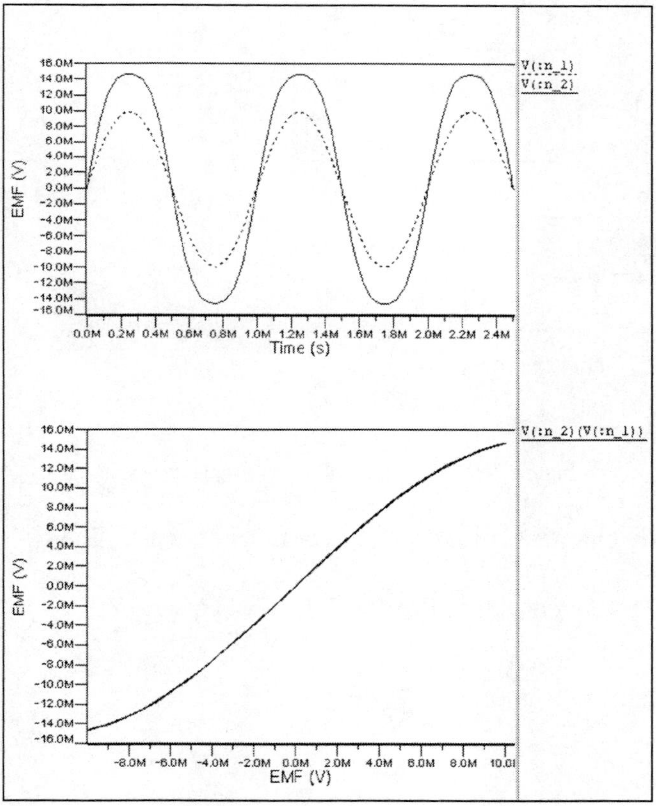

Figure 7-11. Simulation of amplifying a single-tone with nonlinear characteristic

As shown in Figure 7-11 the input signal, which should be doubled in magnitude (power gain of 6dB), is distorted while passing the LNA. The effect of the intercept point IP3 (and finally clipping) is a limitation of the LNA output signal for large input signals. The nonlinear characteristic can be seen in the second diagram.

7.3.2 Mixer

Functional description

A mixer block uses nonlinear circuit characteristics for frequency conversion in RF applications. An analog passband mixer can be realized as a Gilbert cell, which multiplies an RF input signal with a local oscillator (LO) signal to obtain the output signal at an intermediate frequency (IF) together with other spurious signals. This configuration can be found in RF receiver front ends for down conversion. Figure 7-12 shows a simple mixer model, where the Gilbert cell is modeled as a simple multiplier.

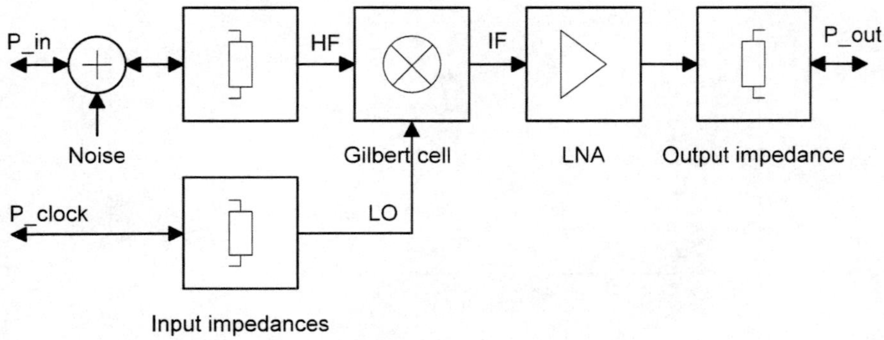

Figure 7-12. Block diagram of a simple mixer model

The main characteristics of the mixer model are as follows.

- LNA characteristics (see Section 7.3.1)

$$v_{out} = a \cdot v_{in} - b \cdot v_{in}^3$$

where

$$a = \sqrt{10^{\frac{gp\_dB}{10}} \cdot \frac{R_{out}}{R_{in}}}, \quad b = \frac{4}{3}\frac{a}{ip3^2}, \quad ip3 = \sqrt{10^{\frac{ip3\_dBm-30}{10}} \cdot 2 \cdot R_{in}}$$

Note: The intercept point in this context refers to the single-tone IP3 of the LNA, while the usually specified IP2 and IP3 for a mixer, which are not parametrizable here, are measured by two tones.

- Simplified Gilbert cell characteristic

$$v_{IF} = v_{HF} \cdot v_{LO}$$

- Input and output impedances are modeled as ohmic resistances. Typical values of a matched system are

$$R_{in} = R_{out} = 50\Omega$$

Model interface

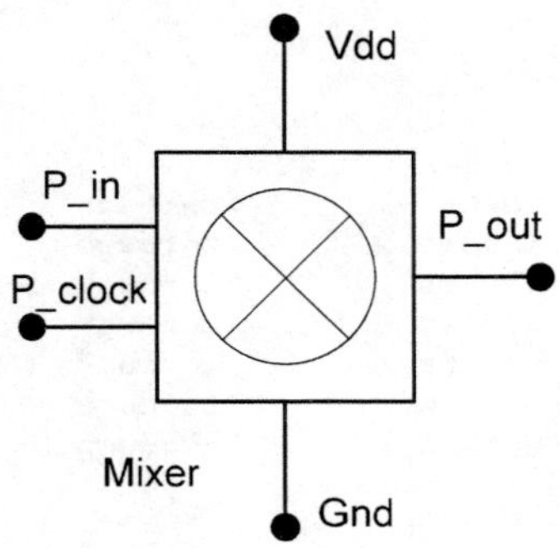

Figure 7-13. Schematic symbol of a mixer

Table 7-13. Model ports

Name	Type	Description
P_IN	ELECTRICAL	Input pin
P_CLOCK	ELECTRICAL	Local oscillator pin
P_OUT	ELECTRICAL	Output pin
VDD	ELECTRICAL	Supply voltage
GND	ELECTRICAL	Reference node

Table 7-14. Model parameters

Name	Unit	Default value	Description
GP_DB	dB	0.0	Open loop power gain
IP3_DBM	dBm	-30.0	Input referenced IP3
FNOISE_DB	dB	0.0	Noise figure of stage
FG	Hz	real'HIGH	Frequency of dominant pole
RIN	Ohm	50.0	Input resistance
RIN_CLK	Ohm	50.0	Clock input resistance
ROUT	Ohm	50.0	Output resistance

Model implementation

```
architecture RF of MIXER is

  constant GP_LIN    : REAL:= 10**(GP_DB/10.0);
                             -- linear value of power gain
  constant IP3_LIN   : REAL:= 10**((IP3_DBM-30.0)/10.0);
                             -- linear value of ip3
  constant A         : REAL:= SQRT(GP_LIN*ROUT/RIN);
                             -- linear value of voltage gain
  constant IP3       : REAL:= SQRT(IP3_LIN*2.0*RIN);
                             -- linear value of ip3 voltage
  constant B         : REAL:= A/(IP3*IP3)*4.0/3.0;
                             -- third order coefficent
  constant INMAX     : REAL:= SQRT(A/(3.0*B));
                             -- maximum input voltage for clipping
  constant OUTMAX    : REAL:= 2.0*A/3.0*INMAX;
                             -- output voltage at clipping
  terminal IN_G      : ELECTRICAL;
  terminal OUT_G     : ELECTRICAL;
  terminal OUT_F     : ELECTRICAL;
  terminal U_OUT     : ELECTRICAL;
  quantity V_NOISE across I_NOISE through P_IN to IN_G;
  quantity V_RIN   across I_RIN   through IN_G to GND;
  quantity V_OUTG  across I_OUTG  through OUT_G to GND;
  quantity V_CLK   across I_CLK   through P_CLOCK to GND;
  quantity V_OUTF  across I_OUTF  through OUT_F to GND;
  quantity V_LIM   across I_LIM   through U_OUT to GND;
  quantity V_ROUT  across I_ROUT  through U_OUT to P_OUT;

begin

  -- input stage: noise figure, input impedances
  V_NOISE == 0.0;
  V_RIN == RIN * I_RIN;
  V_CLK == RIN_CLK * I_CLK;

  -- gilbert cell
  V_OUTG == V_RIN * V_CLK;

  -- lna
  V_OUTF == V_OUTG'LTF((0 => 1.0), (1.0, 1.0/MATH_2_PI/FG));

  if abs(V_OUTF)<INMAX use
    V_LIM == 2.0*(A - B*V_OUTF*V_OUTF)*V_OUTF;
  elsif V_OUTF > 0.0 use
    V_LIM == 2.0*OUTMAX;
  else
    V_LIM == -2.0*OUTMAX;
  end use;

  -- output impedance
  V_ROUT == ROUT * I_ROUT;

end architecture RF;
```

The complete model is included on the CD-ROM that is provided with this book.

Simulation example

For demonstration purposes an RF two-tone signal (1 kHz, -30 dBm and 1.1 kHz, -40 dBm) and a local oscillator of 1 kHz and -30 dBm are used. The carrier is mixed with itself, which is known as a *Direct Conversion* receiver, since no intermediate frequency is used for demodulation.

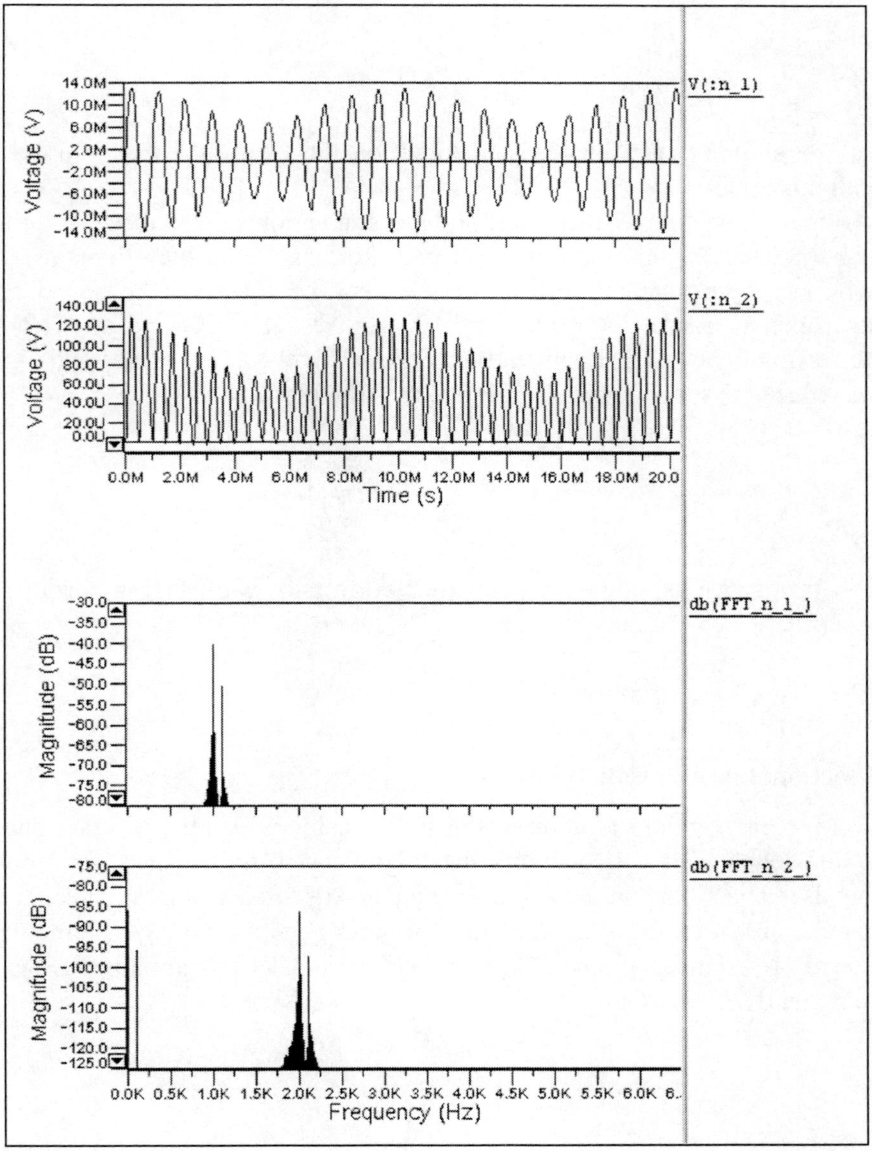

Figure 7-14. Simulation of downconverting (mixing) an RF signal

The mixer model was instantiated with the following parameters:

```
MIX: entity MIXER(RF)
  generic map ( GP_DB => 0.0,
                IP3_DBM => -30.0,
                FNOISE_DB => 2.5
                )
  port map (P_IN => N_1,
            P_OUT => N_2,
            P_CLOCK => N_CLK,
            VDD => N_VDD,
            GND => ELECTRICAL_REF
            );
```

The complete test bench is included on the CD-ROM that is provided with this book.

Figure 7-14 shows the results of the simulation before (node n_1) and after (node n_2) the mixer stage in both time and frequency domains. The two-tone input signal (carrier 1 kHz, signal 1.1 kHz) can be found after down conversion at DC (0 Hz and 100 Hz, respectively) and at twice the carrier frequency (2 kHz and 2.1 kHz, respectively). The carrier and signal are reduced in magnitude according to the formula:

$$a_1 \sin \omega t \cdot a_2 \sin \omega t = \frac{a_1 a_2}{2}\left(1 + \sin\left(2\omega t - \frac{\pi}{2}\right)\right)$$

Other spurious signals appear after conversion well below carrier and signal.

7.3.3 Charge pump

Functional description

The charge pump is an essential building block of mixed-signal phase-locked loops (PLL). It converts two digital pulse trains for up and down signals respectively into analog up and down current pulses. The current sources are built from external positive and negative voltage sources and internal resistances. Figure 7-15 shows the block diagram of the charge pump model.

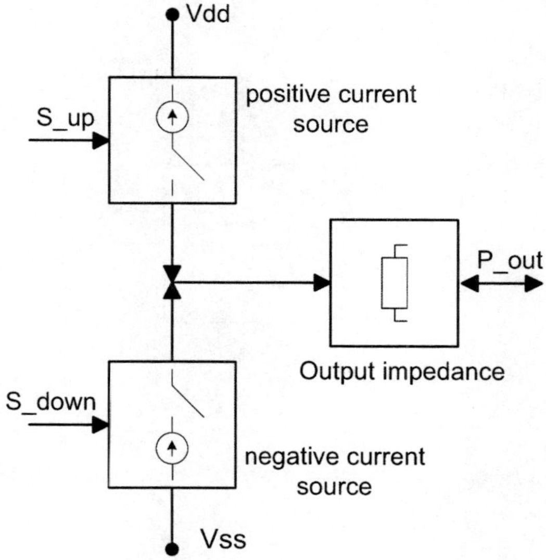

Figure 7-15. Block diagram of a charge pump model

The main characteristics of the charge pump model are as follows.

- Positive and negative current pulses are generated with respective amplitudes

$$I_{pos} = \frac{U_{Vdd}}{R_{isrc}} \text{ and } I_{neg} = \frac{U_{Vss}}{R_{isrc}}$$

where U_{Vdd} and U_{Vss} are the external positive and negative voltage sources, and R_{isrc} is a model parameter.
- The stability of the current can be increased by using high external voltages together with large values for the internal resistors. However, this high voltage may then unwantedly occur at the output of the charge pump when operated without a load.
- In addition, the measurable amplitude of the current pulse at the output pin depends on the specified output resistance and the load resistance, which are both 50 Ω in the often used 50 Ω systems.
- The switches are modeled with two resistances, 1 Ω and 1 MΩ, for the on and off state of the switch, respectively.

Additional characteristics, that were not modeled here, include the explicit voltage limitation of the current sources and nonlinearity.

Model interface

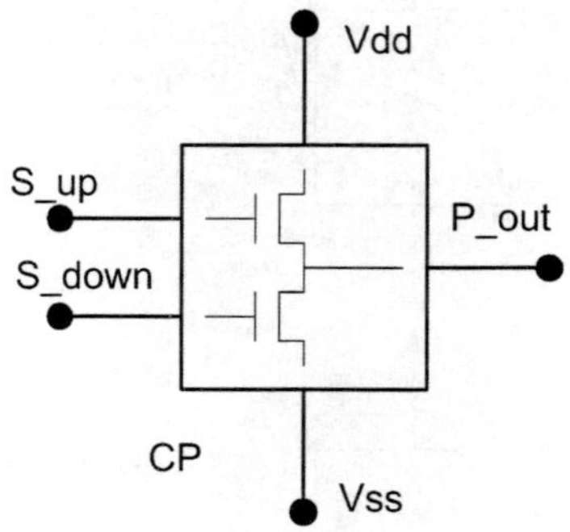

Figure 7-16. Schematic symbol of the charge pump

Table 7-15. Model ports

Name	Type	Description
S_UP	BIT	Up signal
S_DOWN	BIT	Down signal
P_OUT	ELECTRICAL	Output pin
VDD	ELECTRICAL	Positive voltage
VSS	ELECTRICAL	Negative voltage

Table 7-16. Model parameters

Name	Unit	Default value	Description
R_ISRC	Ohm	1.0e3	Current source resistance
ROUT	Ohm	50.0	Output resistance

Model implementation

```
architecture SIMPLE of CP is
  constant RON    : REAL:=1.0;
  constant ROFF   : REAL:=1.0E9;
  terminal N_up   : ELECTRICAL;
  terminal N_down: ELECTRICAL;
  terminal N_int  : ELECTRICAL;
  signal   R_SWUP: REAL:=ROFF;
  signal   R_SWDN: REAL:=ROFF;
  quantity V_SWUP across I_SWUP through N_UP  to N_INT;
  quantity V_SWDN across I_SWDN through N_INT to N_DOWN;
  quantity V_ROUT across I_ROUT through P_OUT to N_INT;
  quantity V_RUP  across I_RUP  through VDD   to N_UP;
  quantity V_RDN  across I_RDN  through VSS   to N_DOWN;
  quantity V_R1   across I_R1   through N_INT to ELECTRICAL_REF;

begin

  R_SWUP <= RON when S_UP='1' else ROFF;
  R_SWDN <= RON when S_DOWN='1' else ROFF;
```

```
    V_SWUP == R_SWUP * I_SWUP;
    V_SWDN == R_SWDN * I_SWDN;
    V_RUP  == R_ISRC * I_RUP;
    V_RDN  == R_ISRC * I_RDN;
    V_ROUT == ROUT * I_ROUT;
    V_R1 == 1.0E6 * I_R1;

  end architecture SIMPLE;
```

The complete model is included on the CD-ROM that is provided with this book.

Simulation example

This example shows the simulation of a digital PLL, which contains, in contrast to the linear PLL, a phase frequency detector and a charge pump. The connected voltage sources have values of +/-5 V.

The charge pump model was instantiated with the following parameters:

```
CHARGEPUMP: entity CP(SIMPLE)
    generic map (R_ISRC => 1.0e6)
    port map (S_UP => S_1,
              S_DOWN => S_2,
              P_OUT => N_TP,
              VDD => P_VDD,
              VSS => N_VSS);
```

The complete test bench is included on the CD-ROM that is provided with this book.

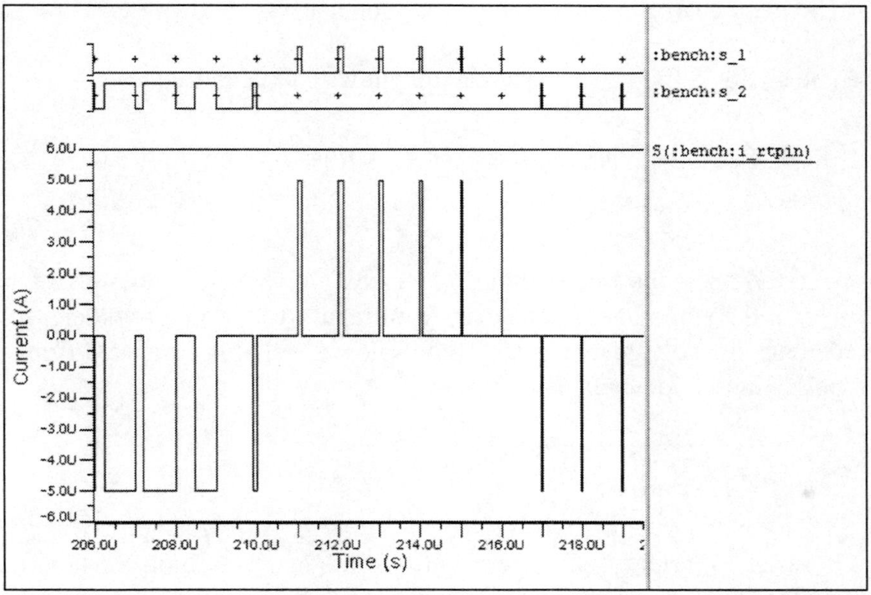

Figure 7-17. Simulation of a charge pump within a digital phase-locked loop (DPLL)

As shown in Figure 7-17, where the output current through the load resistor is shown, the amplitude of the current source equals +/-5 μA according to whether the positive or negative pulse signal is present.

7.3.4 Analog VCO

Functional description

Voltage controlled oscillators (VCO) are used to generate a single-tone sine wave with tunable frequency. Tuning is done using a control voltage at the input pin. Without input voltage the oscillator runs at its free running frequency. Figure 7-18 shows a block diagram of a simple VCO model.

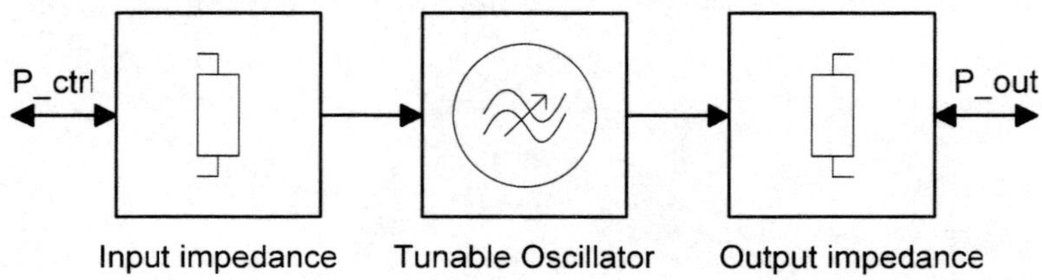

Figure 7-18. Block diagram of a simple VCO model

The main characteristics of the VCO model are as follows.

- The analog VCO has a sinusoidal output with tunable frequency

$$v_{out} = dc + ampl \cdot \sin\left(2\pi \cdot freq_0 \cdot t + \varphi_w + phase_0\right),$$
$$0 = \tfrac{d}{dt}\varphi_w - 2\pi \cdot k_{freq} \cdot v_{in}$$

where $freq_0$ is the free running frequency, $phase_0$ is an initial phase, and k_{freq} is the slope of the frequency with regard to the input voltage. *ampl* denotes the voltage amplitude of the sine wave that is computed from the power amplitude *ampl_dBm* by

$$ampl = \sqrt{2 \cdot rout \cdot 10^{\frac{ampl\_dBm-30}{10}}}$$

In order to ensure a consistent initial solution a condition for the phase angle during the operating point analysis has been added

$\varphi_w = 0$ for $t = 0$.

- Input and output impedances are modeled as ohmic resistances. Typical values of a matched system are:

$$R_{in} = R_{out} = 50\Omega$$

Additional characteristics, that were not modeled here, include

- Timing jitter and phase noise (see [Lee01])
- Nonlinear tuning characteristic
- Second order effects such as temperature and power supply dependency of the tuning characteristic
- Dead time of the control input
- Power consumption

Model interface

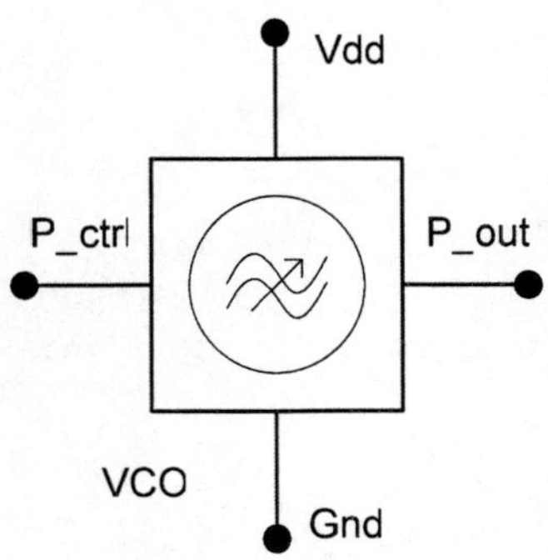

Figure 7-19. Schematic symbol of VCO

Table 7-17. Model ports

Name	Type	Description
P_CTRL	ELECTRICAL	Control pin
P_OUT	ELECTRICAL	Output pin
VDD	ELECTRICAL	Supply voltage
GND	ELECTRICAL	Reference node

Table 7-18. Model parameters

Name	Unit	Default value	Description
DC	V	0.0	DC offset voltage of oscillator
AMPL_DBM	dBm	-100.0	Amplitude of sine wave
FREQ_0	Hz	1.0e+03	Free running oscillator frequency
PHASE_0	rad	0.0	Constant phase offset
K_FREQ	Hz/V	1.0e+06	Sensitivity of voltage input
RIN	Ohm	50.0	Input resistance
ROUT	Ohm	50.0	Output resistance

Model implementation

```
architecture ANALOG of VCO is
  constant AMPL_LIN : REAL:= 10**((AMPL_DBM-30.0)/10.0);
  constant AMPL     : REAL:= SQRT(AMPL_LIN * 2.0 * ROUT);
  terminal N_INT    : ELECTRICAL;
  quantity V_RIN  across I_RIN  through P_CTRL to GND;
  quantity V_ROUT across I_ROUT through P_OUT to N_INT;
  quantity V_SRC  across I_SRC  through N_INT to GND;
  quantity PHI_W : REAL;
begin
-- input impedance
  V_RIN  == RIN * I_RIN;

-- tunable oscillator
  if DOMAIN = QUIESCENT_DOMAIN use
    PHI_W == 0.0;
  else
    0.0 == PHI_W'DOT - MATH_2_PI*K_FREQ*V_RIN;
  end use;

  V_SRC == 2.0 * (DC + AMPL * SIN(NOW * MATH_2_PI*FREQ_0
                                 + PHI_W + PHASE_0));
-- output impedance
  V_ROUT == ROUT * I_ROUT;

end architecture ANALOG;
```

The complete model is included on the CD-ROM that is provided with this book.

Simulation example

A step signal was applied as the control voltage, which changes at 10.2 ms from 0 to –5 mV. The VCO model was instantiated with the following parameters:

```
LO: entity VCO(ANALOG)
      generic map ( AMPL_DBM => -30.0,
                    FREQ_0 => 1.0e03,
                    K_FREQ => 1.0e05)
      port map (P_CTRL => N_CTRL,
                P_OUT => N_CLK,
                VDD => N_VDD,
                GND => ELECTRICAL_REF);
```

The complete test bench is included on the CD-ROM that is provided with this book.

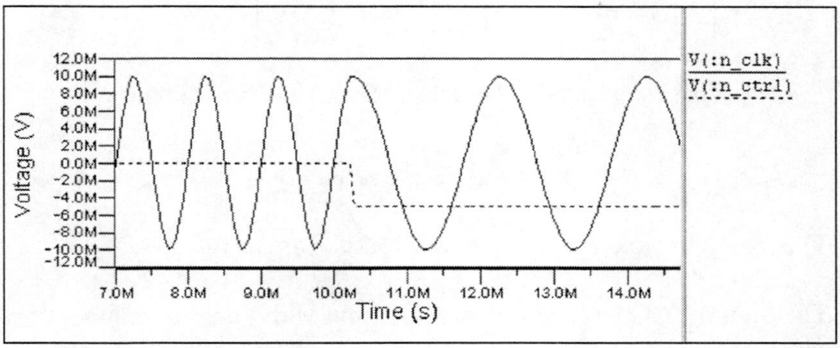

Figure 7-20. Simulation of frequency changes when altering the control voltage

Figure 7-20 shows the simulation results. When the control voltage is zero, the output oscillates at the free running frequency of 1 kHz. At time 10.2 ms a negative control voltage of –5 mV is applied. This leads to half the original frequency

$$f_{out} = f_0 + k_f \cdot v_{in} = 1 \cdot 10^3 - 1 \cdot 10^5 \cdot 5 \cdot 10^{-3} = 500 \text{ Hz}$$

Although the frequency does not change at a zero crossing the phase continuously changes.

7.3.5 Digital VCO

Functional description

Digital voltage controlled oscillators (VCO) are used to generate a single-tone square wave with tunable frequency. Tuning is done using a control voltage at the input pin. Without input voltage the oscillator runs at its free running frequency. Figure 7-21 shows a block diagram of a simple VCO model.

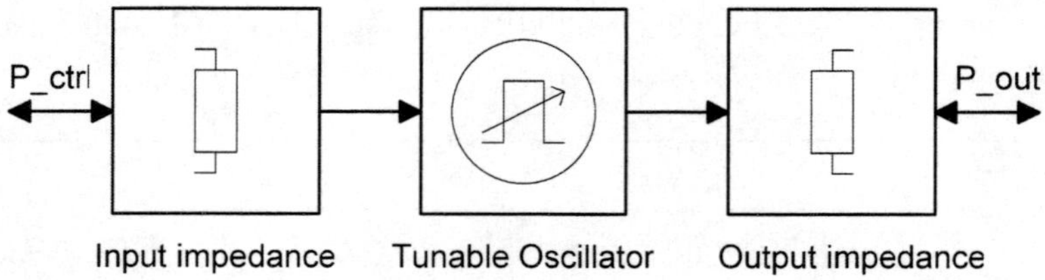

Figure 7-21. Block diagram of a simple digital VCO model

The main characteristics of the VCO model are as follows.

- The digital VCO has a square wave output with tunable frequency

  $$v_{out} = dc \pm ampl, \;\; f = freq_0 + k_{freq} \cdot v_{in}$$

  where $freq_0$ is the free running frequency, k_{freq} is the slope of the frequency with regard to the control voltage. *ampl* denotes the voltage amplitude of the sine wave that is computed from the power amplitude *ampl_dBm* by

  $$ampl = \sqrt{2 \cdot \boldsymbol{rout} \cdot 10^{\frac{ampl\_dBm-30}{10}}}$$

- For a nonpositive frequency f the model stops oscillating until $f > 0$ is satisfied.

- In this simple model the frequency f can only change at transition times, that is at a rising or falling edge. For more accurate modeling the frequency should be changeable all the time, which means that the scheduling time of the subsequent edge has to be permanently adjusted.
- Input and output impedances are modeled as ohmic resistances. Typical values of a matched system are

  $$R_{in} = R_{out} = 50\Omega$$

Additional characteristics, that were not modeled here, include

- Transition times for rising and falling edge
- Timing jitter and phase noise (see [Lee01])
- Nonlinear tuning characteristic

- Second order effects such as temperature and power supply dependency of the tuning characteristic
- Dead time of the control input
- Power consumption

Model interface

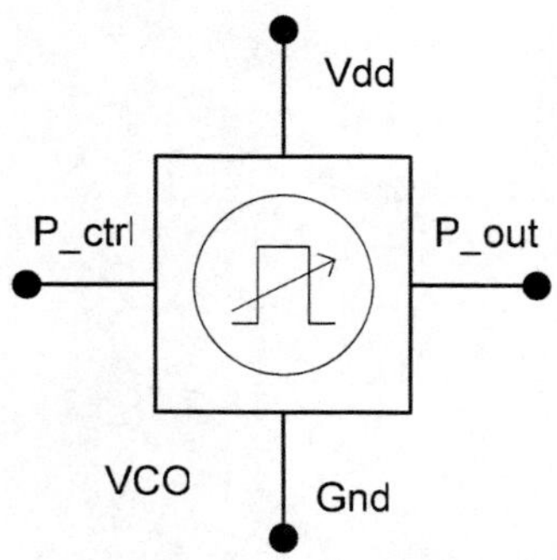

Figure 7-22. Schematic symbol of VCO

Table 7-19. Model ports

Name	Type	Description
P_CTRL	ELECTRICAL	Control pin
P_OUT	ELECTRICAL	Output pin
VDD	ELECTRICAL	Supply voltage
GND	ELECTRICAL	Reference node

Table 7-20. Model parameters

Name	Unit	Default value	Description
DC	V	0.0	DC offset voltage of oscillator
AMPL_DBM	dBm	-100.0	Amplitude of sine wave
FREQ_0	Hz	1.0e+03	Free running oscillator frequency
K_FREQ	Hz/V	1.0e+06	Sensitivity of voltage input
RIN	Ohm	50.0	Input resistance
ROUT	Ohm	50.0	Output resistance

Model implementation

```
architecture DIGITAL of VCO is
  constant AMPL_LIN : REAL:= 10**((AMPL_DBM-30.0)/10.0);
  constant AMPL     : REAL:= SQRT(AMPL_LIN * 2.0 * ROUT);
  signal   STATE    : bit:='0';
  signal   FACTOR   : real:=0.0;
  terminal N_INT    : ELECTRICAL;
```

```
    quantity V_RIN  across I_RIN  through P_CTRL to GND;
    quantity V_ROUT across I_ROUT through P_OUT to N_INT;
    quantity V_SRC  across I_SRC  through N_INT to GND;
  begin

    process (STATE,V_RIN'ABOVE(-FREQ_0/K_FREQ))
      VARIABLE F: REAL;
    begin
      if STATE='1' THEN
        FACTOR<=1.0;
      else
        FACTOR<=-1.0;
      end if;
      F := FREQ_0 + K_FREQ*V_RIN;
      if F>0.0 then
        STATE <= not STATE after 0.5/F*SEC;
      end if;
    end process;
    break on STATE;

  -- input impedance
    V_RIN  == RIN * I_RIN;

  -- tunable oscillator
    V_SRC == 2.0*(DC+FACTOR*AMPL);

  -- output impedance
    V_ROUT == ROUT * I_ROUT;

  end architecture digital;
```

The complete model is included on the CD-ROM that is provided with this book.

Simulation example

A piecewise linear signal was applied as the control voltage, which changes at 10.2 ms from 0 to –15 mV, and at 12.2 ms to –5 mV. The VCO model was instantiated with the following parameters:

```
LO: entity VCO(DIGITAL)
      generic map ( AMPL_DBM => -30.0,
                    FREQ_0 => 1.0e03,
                    K_FREQ => 1.0e05)
      port map (P_CTRL => N_CTRL,
                P_OUT => N_CLK,
                VDD => N_VDD,
                GND => ELECTRICAL_REF);
```

The complete test bench is included on the CD-ROM that is provided with this book.

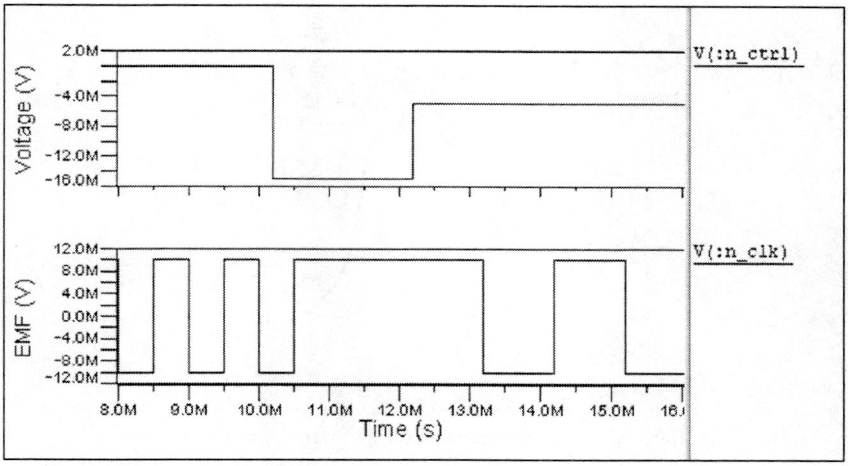

Figure 7-23. Simulation of frequency changes when altering the control voltage

Figure 7-23 shows the simulation results. When the control voltage is zero, the output oscillates at the free running frequency of 1 kHz. At time 10.2 ms a negative control voltage of –15 mV is applied. This leads to a negative frequency of

$$f_{out} = f_0 + k_f \cdot v_{in} = 1 \cdot 10^3 - 1 \cdot 10^5 \cdot 15 \cdot 10^{-3} = -500\,\mathrm{Hz}$$

Therefore the oscillation stops after the next transition. At 12.2 ms a voltage of 5 mV is applied to the control input, which leads to a frequency of

$$f_{out} = f_0 + k_f \cdot v_{in} = 1 \cdot 10^3 - 1 \cdot 10^5 \cdot 5 \cdot 10^{-3} = 500\,\mathrm{Hz}$$

Thus, the oscillator restarts with a halfperiod of 1 ms, that is a falling edge at approximately 13.2 ms.

7.3.6 Filters

Functional description

Filters are mainly characterized by their behavior in the frequency domain. Some frequencies of the input signal pass through the filter (passband), while others are rejected (stopband). Real filters have, in contrast to ideal filters, a transition band in between, where the frequency response changes continuously from passing to rejection (see Figure 7-24).

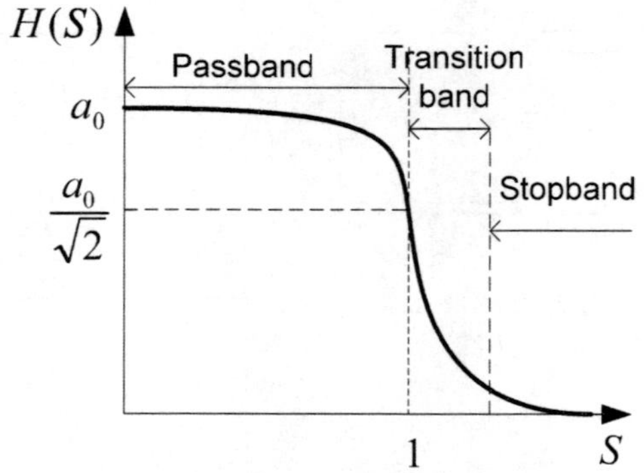

Figure 7-24. Frequency response of a real lowpass filter

We may distinguish between highpass, lowpass, bandpass and bandstop filters. Depending on the type of polynomial used in the transition function *H(s)*, different types of filters are available such as Chebyshev, Butterworth, elliptical, etc. The frequency responses differ in the tolerances of passband, stopband and the slope of the transition. Other categories of filters are possible, for example analog or digital, finite or infinite impulse response (FIR and IIR). A general block diagram is shown in Figure 7-25.

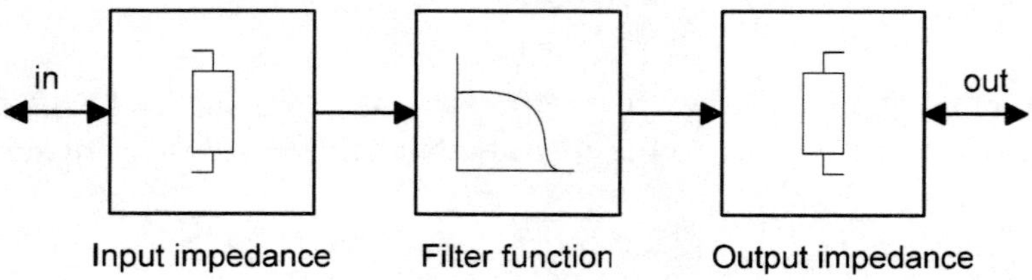

Figure 7-25. Block diagram of an RF filter model

In this section we only consider Butterworth type filters, which are known to have a maximally flat frequency response. For a lowpass filter this is expressed in the Laplace domain as

$$H(S) = \frac{a_0}{\prod_i (1 + a_i \cdot S + b_i \cdot S^2)}, \text{ with } i = 1, \ldots, \; S = \frac{s}{2\pi f_g},$$

where S is the complex variable normalized on the frequency f_g.

Through transformation of the variables a highpass description can be obtained from this lowpass transfer function.

The main characteristics of the lowpass filter are as follows:

- The filter gain is measured at zero frequency

$$a_0 = \sqrt{10^{\frac{gain}{10}} \frac{R_{out}}{R_{in}}}$$

- The grade n of filter, which is the exponent of the highest power of S in $H(S)$ and determines the range of the transition band as well as the attenuation in the stopband.
Note: In our implementation the grade n of the filter is limited to 6, since a more general model using the GENERATE construct of VHDL-AMS is not yet supported by the simulator.
- Coefficients of the transition function $H(S)$ are for even values of grade n

$$a_i = 2 \cdot \cos\frac{(2i-1)\cdot\pi}{2n},\ b_i = 1,\ i = 1,\ldots,\frac{n}{2}$$

and for odd values of grade n

$$a_1 = 1,\ a_i = 2 \cdot \cos\frac{(i-1)\cdot\pi}{n},$$

$$b_1 = 0,\ b_i = 1,\ i = 2,\ldots,\frac{n+1}{2}$$

- The cut-off frequency f_g (or 3dB frequency) is measured where the gain drops to

$$\frac{a_0}{\sqrt{2}}$$

- Input and output impedances are modeled as ohmic resistances. Typical values of a matched system are: $R_{in} = R_{out} = 50\Omega$

Additional characteristics, that were not modeled here, include second order effects such as temperature and power supply dependency of the filter characteristic, and power consumption.

Model interface

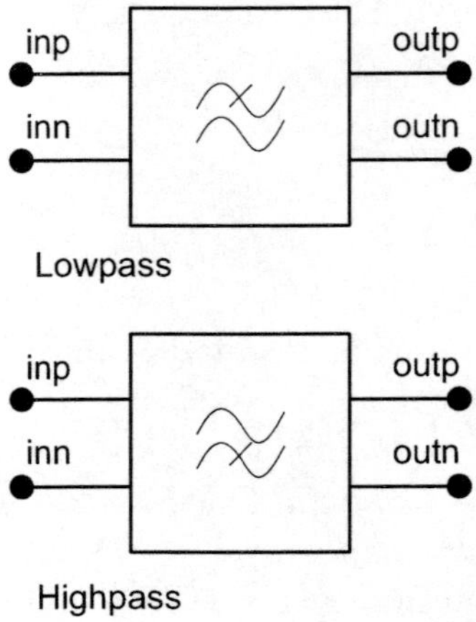

Figure 7-26. Schematic symbols of different filters

Table 7-21. Model ports

Name	Type	Description
INP	ELECTRICAL	Positive input pin
INN	ELECTRICAL	Negative input pin
OUTP	ELECTRICAL	Positive output pin
OUTN	ELECTRICAL	Negative output pin

Table 7-22. Model parameters

Name	Unit	Default value	Description
GAIN	dB	0.0	Maximum power gain of amplifier
FG	Hz	1.0	Cut-off frequency
GRAD		1	Grade of filter
RIN	Ohm	50.0	Input resistance
ROUT	Ohm	50.0	Output resistance

Model implementation

The filter coefficients are precomputed within a package and given to the model as a vector.

```
function LOWPASS_BUTTERWORTH_A (GRAD : INTEGER)
  return REAL_VECTOR is
  constant NU   : INTEGER := (GRAD+1)/2;
  variable A    : REAL_VECTOR (1 to NU);
begin
  if GRAD mod 2 = 0 then
    for I in 1 to GRAD/2 loop
```

```
      A(I) := 2.0*COS((2.0*REAL(I)-1.0)*MATH_PI/2.0/REAL(GRAD));
    end loop;
  else
    A(1) := 1.0;
    for I in 2 to (GRAD+1)/2 loop
      A(I) := 2.0*COS((REAL(I)-1.0)*MATH_PI/REAL(GRAD));
    end loop;
  end if;
  return A;
end function LOWPASS_BUTTERWORTH_A;

function LOWPASS_BUTTERWORTH_B (GRAD : INTEGER)
  return REAL_VECTOR is
  constant NU  : INTEGER := (GRAD+1)/2;
  variable B   : REAL_VECTOR (1 to NU);
begin
  if GRAD mod 2 = 0 then
    for I in 1 to GRAD/2 loop
      B(I) := 1.0;
    end loop;
  else
    B(1) := 0.0;
    for I in 2 to (GRAD+1)/2 loop
      B(I) := 1.0;
    end loop;
  end if;
  return B;
end function LOWPASS_BUTTERWORTH_B;
```

The model implementation is included on the CD-ROM that is provided with this book.

Simulation example

The simulation is performed with six lowpass filters in parallel, which have an increasing filter grade from one to six. A step signal of amplitude one is applied to all filter blocks. The model call for the filter with degree one looks like:

```
UUT1: entity LOWPASS_FILTER (BHV_RF)
        generic map (GAIN, FG, 1)
        port map (I1, ELECTRICAL_REF, O1, ELECTRICAL_REF);
```

The complete test bench is included on the CD-ROM that is provided with this book.

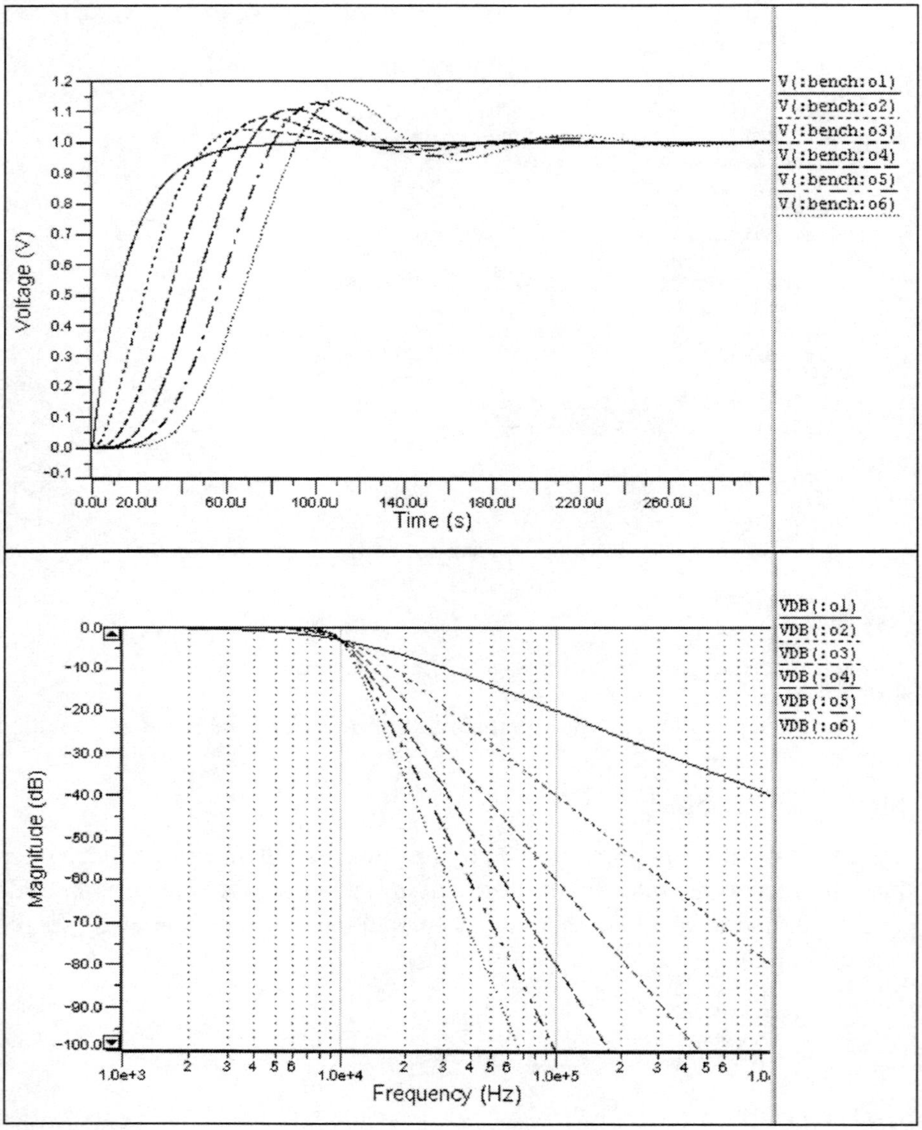

***Figure** 7-27.* Simulation of filter responses in time and frequency domain

Figure 7-27 shows the simulation results. In the frequency domain it can be seen how the attenuation in the stop-band increases with the filter grade ($n \cdot 20\text{dB/decade}$, n…filter grade). The attenuation of 3 dB compared to a_0 at the cut-off frequency is common to all six curves.

In the time domain, filters with a higher grade show a slower step response and more overshooting.

7.3.7 Switch

Functional description

Opening or closing a branch in a network always requires a structural change in the describing system of equations. Therefore these changes are difficult to model and to handle by the simulator. Here we use a simple modeling method where only the resistance of a network branch changes from a very high to a very low value.

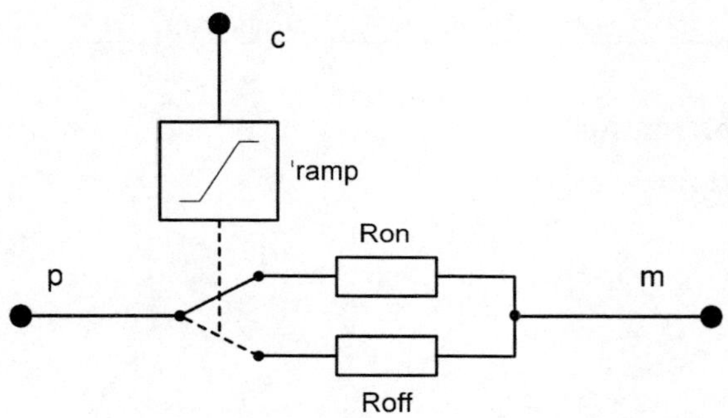

Figure 7-28. Block diagram of the switch model

The main characteristics of the switch model are as follows.

- Switching between two adjustable resistances *Ron* and *Roff* according to the control signal.
- Adjustable transition time for on and off switching (*ton* and *toff*) using the *'ramp* construct.

Further refined models would include a nonlinear transition behavior instead of a piecewise linear characteristic.

Model interface

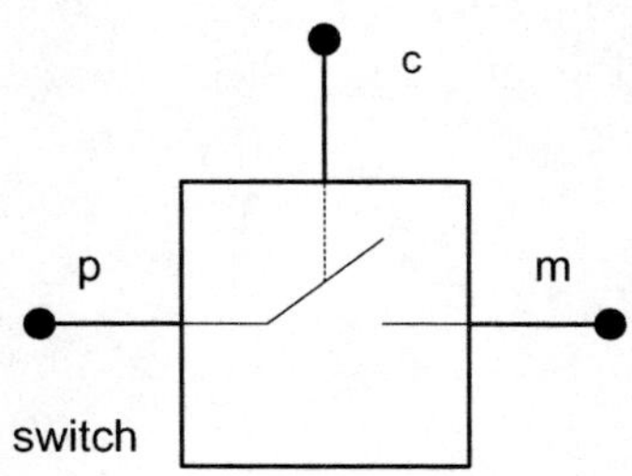

Figure 7-29. Schematic symbol of a switch

Table 7-23. Model ports

Name	Type	Description
P	ELECTRICAL	Positive pin
M	ELECTRICAL	Negative pin
C	in STD_LOGIC	Control port

Table 7-24. Model parameters

Name	Unit	Default value	Description
RON	Ohm	1.0e-03	Resistance when switch is closed
ROFF	Ohm	1.0e06	Resistance when switch is open
TON	s	1.0e-06	Transition time for rising edge (off to on)
TOFF	s	1.0e-06	Transition time for falling edge (on to off)

Model implementation

```
architecture RAMP of SWITCH is

  signal    R_VAL : REAL := ROFF;
  quantity V_SW across I_SW through P    to M;

begin

  R_VAL <= RON  when C = '1' else
           ROFF when C = '0';

  V_SW == R_VAL'RAMP(TON, TOFF)*I_SW;

end architecture RAMP;
```

The complete model is included on the CD-ROM that is provided with this book.

7.3.8 General n-bit A/D and D/A converter

Functional description

An analog to digital and a digital to analog converter are presented with a general n-bit vector interface at the digital side. The equivalent analog power is parameterizable by a model parameter. The A/D converter is triggered with an additional clock input. Figure 7-30 shows the block diagrams of both the A/D and the D/A converter.

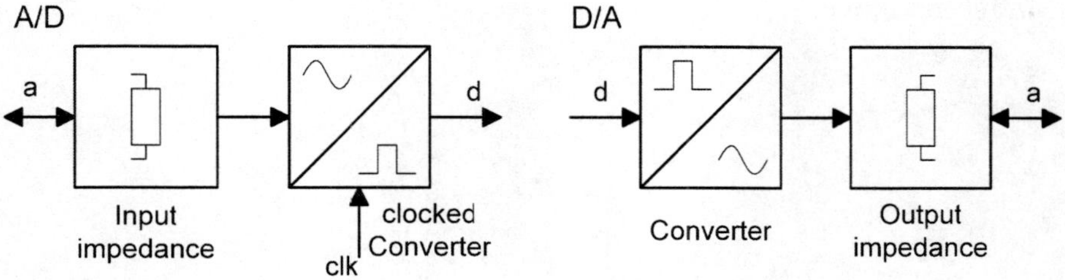

Figure 7-30. Block diagram of A/D and D/A converter models

The main characteristics of the converter models are as follows.

- Bit vector input has arbitrary width.
- Maximum power at the analog pin can be parameterized by *pmax_dBm* and leads to a maximum voltage of

$$va = \sqrt{2 \cdot rout \cdot 10^{\frac{pmax\_dBm-30}{10}}}$$

- When a bit vector of width *d'length* is used, the maximum power is subdivided into $2^{d'length} - 1$ steps, so the least significant bit corresponds to a voltage of

$$v_{lsb} = \frac{va}{2^{d'length} - 1}$$

- Transition times for rising and falling edge are included.
- The output impedance is modeled as an ohmic resistance

Additional characteristics, that were not modeled here, include

- Dead time and hysteresis for conversion
- Nonlinear conversion characteristic
- Power consumption

Model interface

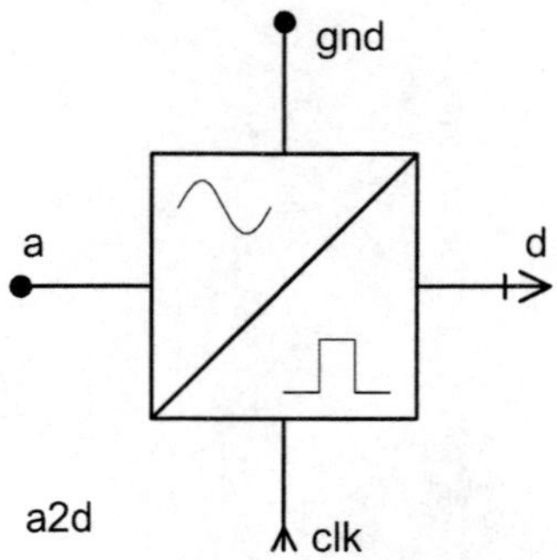

Figure 7-31. Schematic symbol of A/D converter

Table 7-25. Model ports of A/D converter

Name	Type	Description
A	ELECTRICAL	Analog input
GND	ELECTRICAL	Analog ground
D	out BIT_VECTOR	Digital output
CLK	in BIT	Clock input

Table 7-26. Model parameters of A/D converter

Name	Unit	Default value	Description
PMAX_DBM	dBm	-100.0	Maximum power amplitude
RIN	Ohm	50.0	Input resistance

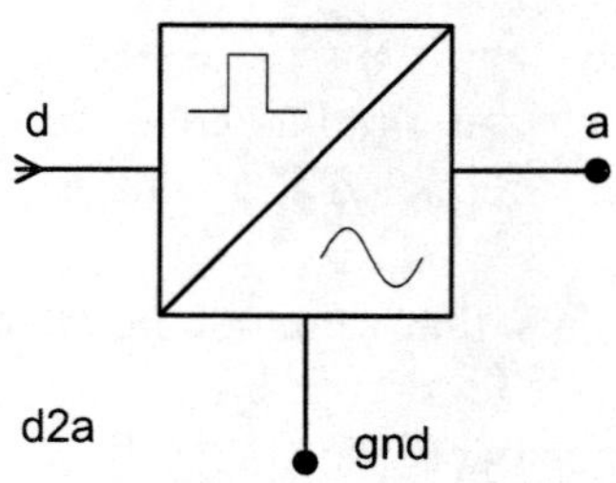

Figure 7-32. Schematic symbol of D/A converter

Table 7-27. Model ports of A/D converter

Name	Type	Description
D	in bit_vector	Digital input
A	ELECTRICAL	Analog output
GND	ELECTRICAL	Analog ground

Table 7-28. Model parameters of A/D converter

Name	Unit	Default value	Description
PMAX_DBM	dBm	-100.0	Maximum power amplitude
T_RISE	s	1.0e-09	Transition time for rising edge
T_FALL	s	1.0e-09	Transition time for falling edge
ROUT	Ohm	50.0	Output resistance

Model implementation

The VHDL-AMS code of the digital process for D/A conversion is given.

```
-- conversion process
process (D) is
  variable NUMBER   : INTEGER;
  begin
    NUMBER := 0;
    for I in D'HIGH downto D'LOW loop
      NUMBER := 2 * NUMBER;
      if D(I) = '1' then
        NUMBER := NUMBER + 1;
      end if;
    end loop;

    VALUE <= REAL(NUMBER)*VSB;
end process;
```

The complete source code of A/D and D/A converter models are included on the CD-ROM that is provided with this book.

Simulation example

A pseudorandom binary source was used as the input for the digital to analog converter. It has been parameterized with a generator polynomial length of six, corresponding to a sequence period of 127 bits, and an output frequency of 5 MHz, corresponding to a bit time of 200 ns. The source and the D/A converter were instantiated with the following parameters.

```
SRC3: entity PRBS(SHIFT_REGISTER)
        generic map (POLYGRAD => 6,
                     BIT_TIME => 1.0/(5.0e06)*1sec)
        port map (BIT_OUT => S_BIT(0));

CONV: entity D2A(BHV_RF)
        generic map(PMAX_DBM => -30.0)
        port map (D => S_BIT,
                  A => N_A,
                  GND => ELECTRICAL_REF);
```

The complete test-bench is included on the CD-ROM that is provided with this book.

The D/A converter computes a new output voltage as the input bitstream changes. The maximum output power value of –30 dBm corresponds to a voltage of 10 mV at the standard terminating resistor of 50 Ω.

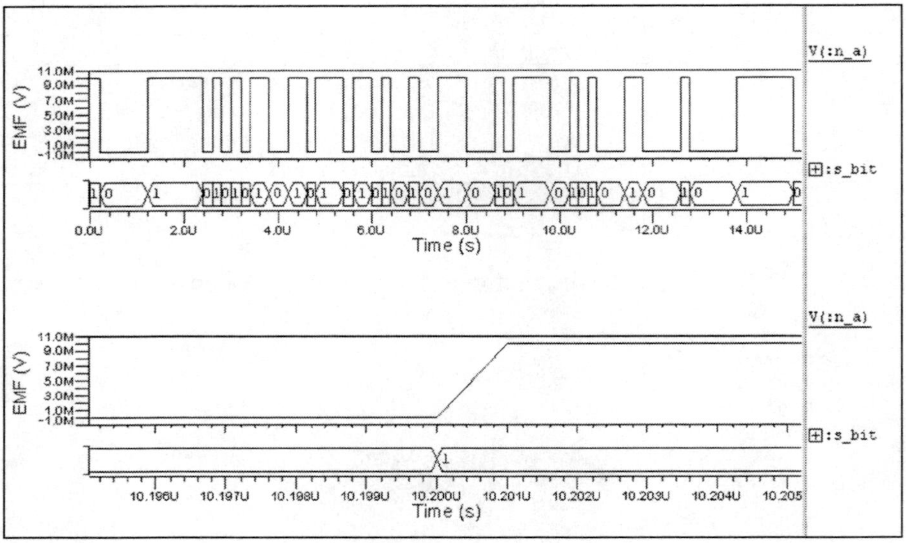

Figure 7-33. Simulation of the D/A conversion process

Figure 7-33 shows the simulation results in the analog and the digital domain. In the upper half of the results window the whole sequence of 150 bits over 30 µs is depicted. The lower half shows a zoom view of the region of 20 ns. There is to be seen that after a change on the digital side the analog voltage immediately follows with a rising edge of 1 ns duration (default value).

A second example was created to test the A/D converter in series with the D/A converter. Therefore as the input signal a sine wave is first sampled and converted to a digital signal. Then the sine wave is reconstructed by converting the digital signal back to the analog domain. The converters were instantiated with the following parameters.

```
A2DC: entity A2D(BHV_RF)

        generic map(PMAX_DBM => -30.0)
        port map  (A => N_A,
                   GND => ELECTRICAL_REF,
                   D => S_BIT,
                   CLK => S_CLK);

D2AC: entity D2A(BHV_RF)

        generic map(PMAX_DBM => -30.0,
                     TRISE => 1.0E-10,
                     TFALL => 1.0E-10)
        port map  (D => S_BIT,
                   A => N_A2,
                   GND => ELECTRICAL_REF);
```

The complete test-bench is included on the CD-ROM that is provided with this book.

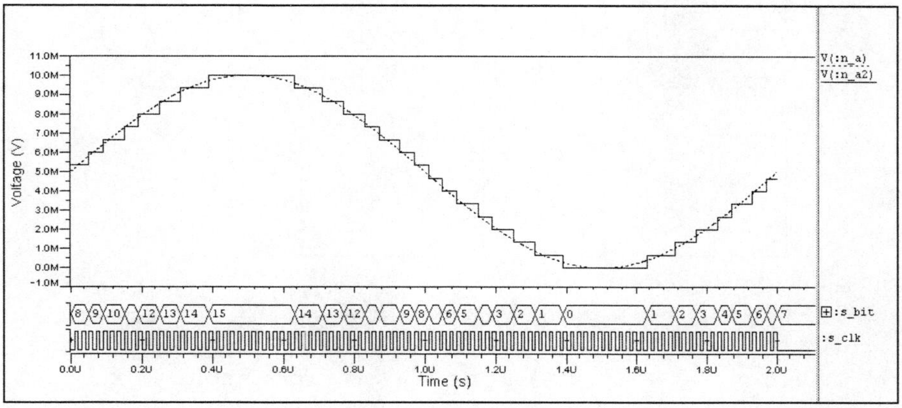

Figure 7-34. Simulation of sampling and reconstructing a sine wave

The results window in Figure 7-34 shows the analog input waveform at node *n_a*. It has a frequency of 50 kHz. It is sampled by the A/D converter with a clock signal of 5 MHz. The 4-bit vector with values between 0 and 15 is then fed to the D/A converter. The analog result is displayed as a voltage at node *n_a2* and follows the original sine wave.

7.3.9 Simple channel

Functional description

In system level simulation, channel models often comprise all the effects that occur on a signal between the analog front-ends of a transmitter and receiver. The channel model described here only includes the basic effects of additive white Gaussian noise (AWGN) and delay. Figure 7-35 shows the block diagram of a simple channel model.

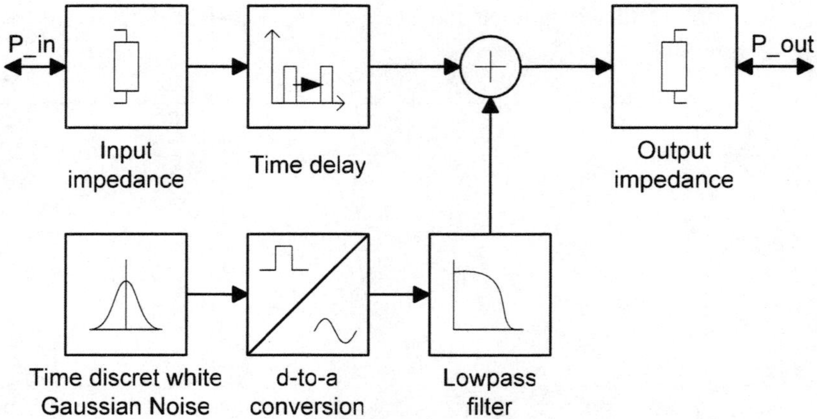

Figure 7-35. Block diagram of a simple channel model

The main characteristics of the simple channel model are as follows.

- Additive white Gaussian noise (AWGN) with power level *pn_dBm*, which is specified by the input signal power *ps_dBm* and the signal-to-noise ratio *s_to_n*

$$vn = \sqrt{10^{\frac{pn\_dBm-30}{10}} \cdot Z_0}\text{, where } pn\_dBm = ps\_dBm - s\_to\_n$$

$v_{noise} = vn \cdot \sqrt{2 \cdot \log(x_1^{-1})} \cdot \cos(2\pi x_2)$, where *x1*, *x2* are uniformly distributed random numbers ($0 \le x_{1,2} \le 1,\quad x_1 \ne 0$)

- For the input signal a delay *td* can be specified.
- Input and output impedances are modeled as ohmic resistances with default values

$$Z_0 = R_{in} = R_{out} = 50\Omega$$

Additional characteristics, that were not modeled here, include multi-path propagation of the signal due to reflection with different delays and damping factors for different paths, frequency response, nonlinear distortion, etc. Thus far the model is very general and suited to RF applications as well as to wired data transmission, where effects like crosstalk and reflection due to mismatch are predominant.

Model interface

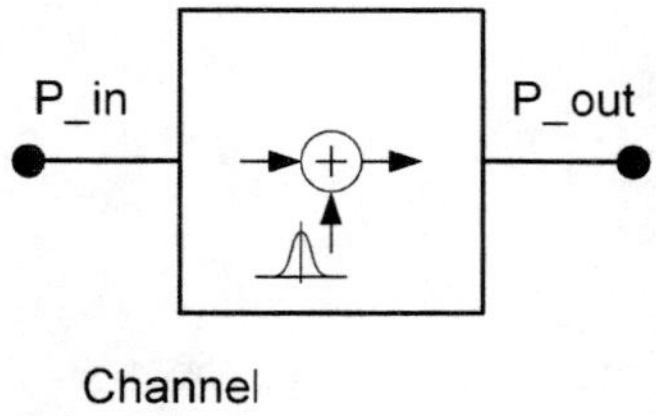

Figure 7-36. Schematic symbol of AWGN channel

Table 7-29. Model ports

Name	Type	Description
P_IN	ELECTRICAL	Connection pin
P_OUT	ELECTRICAL	Connection pin

Table 7-30. Model parameters

Name	Unit	Default value	Description
PS_DBM	dBm	-100.0	Input signal power in dBm
S_TO_N	dB	100.0	Signal-to-noise ratio
FS_NOISE	Hz	1.0e06	Sampling frequency of noise signal
TD	s	0.0	Time delay for input signal
RIN	Ohm	50.0	Input resistance
ROUT	Ohm	50.0	Output resistance

Model implementation

The core of the model is the following digital process that generates time discrete white Gaussian noise.

```
process is
  variable X1,X2,X: REAL := 0.0;
  variable SD1: POSITIVE := 111;
  variable SD2: POSITIVE := 333;
begin
  UNIFORM(SD1, SD2, X1);      -- uniform gives a value 0<x<1
  UNIFORM(SD1, SD2, X2);      -- defined in ieee.math_real
  X:=VN*COS(MATH_2_PI*X1)*SQRT(-2.0*LOG(X2));
  S_NOISE<=X;
  wait for PERIOD;
end process;
```

The complete model is included on the CD-ROM that is provided with this book.

Simulation example

In the simulation example a single-tone of 1 kHz and –30 dBm was fed to the channel model, which is parameterized with a signal-to-noise ratio of

20 dB and a sampling frequency for the noise signal of 1 MHz. The channel model was instantiated with the following parameters.

```
CHN: entity CHANNEL(AWGN)
       generic map (PS_DBM => -30.0,
                    S_TO_N => 20.0,
                    FS_NOISE => 1.0e06)
       port map (P_IN=>N_1, P_OUT=>N_2);
```

The complete test-bench is included on the CD-ROM that is provided with this book.

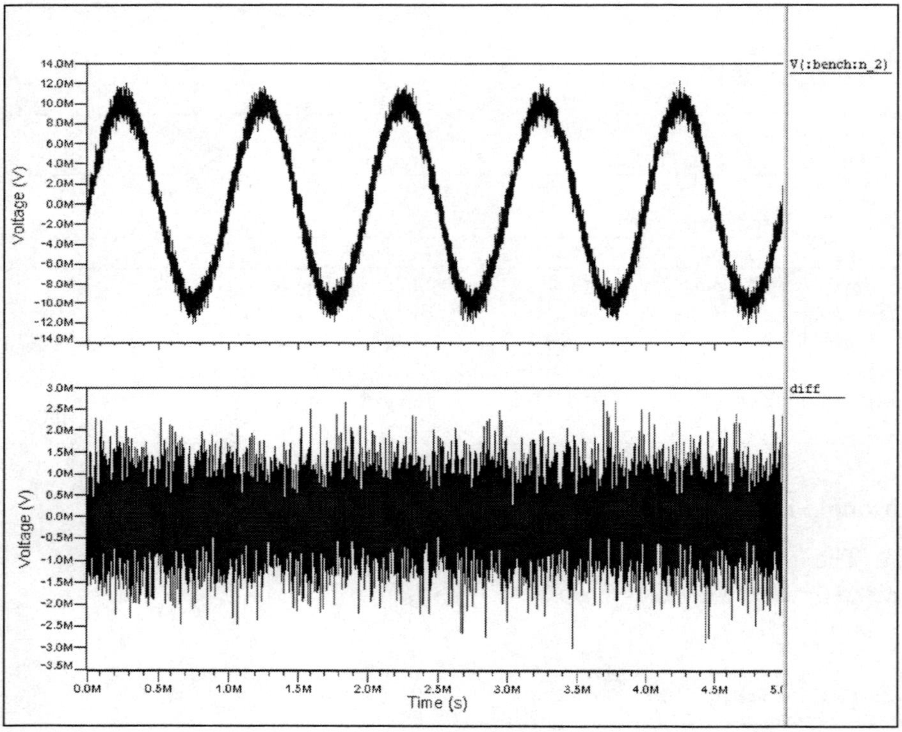

Figure 7-37. Results of the AWGN channel simulation in time domain

Figure 7-37 shows the simulation results. The input signal is superimposed by a noisy signal, which is depicted as the difference *(diff)* between the input and the output signal (*V:bench:n_2*).

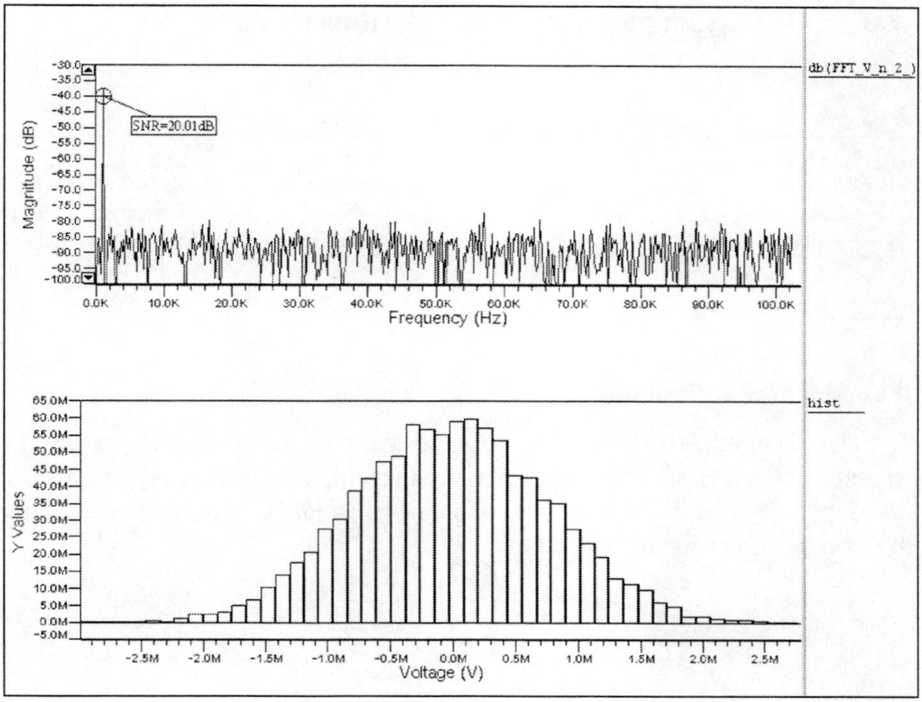

Figure 7-38. Time domain analysis of the noise signal

In the frequency domain (upper part of Figure 7-38) a peak of –40 dB at 1 kHz is seen, which corresponds to input power of –30 dBm. The noise floor is around –90 dB and "white" up to 1 MHz (although it is only displayed to 100 kHz to allow visibility of the input signal at 1 kHz). In addition the analysis of the signal-to-noise ratio is shown. The integration yields

$$\mathrm{SNR} = \frac{\int_{\mathrm{Fmin}}^{\mathrm{Fmax}} S_S(f)df}{\int_{\mathrm{Fmin}}^{\mathrm{Fmax}} S_N(f)df} = 20.01\,\mathrm{dB}$$

where $S_X(f)$ is the power spectral density (PSD) of signal X. This value matches the chosen parameter value S_TO_N=20 dB.

From the difference signal in the time domain a histogram is computed using the post processing capabilities of the waveform tool (lower part of Figure 7-38). In the histogram the Gaussian nature of that noise can be observed.

7.4 Measurement and Observation Units

Table 7-31. Overview on measurement and observation units

Model	Properties
Peak detector	Simple model using 'SLEW attribute
Frequency measurement unit	Frequency measurement for periodic signals
Power meter	Logarithmic measure of the average power

7.4.1 Peak detector

Functional description

The function of this block is to keep track of the maximum value of a signal. Whenever the input signal rises the output immediately follows. If the input signal falls, the output remains constant. Figure 7-39 shows the block diagram of the peak detector.

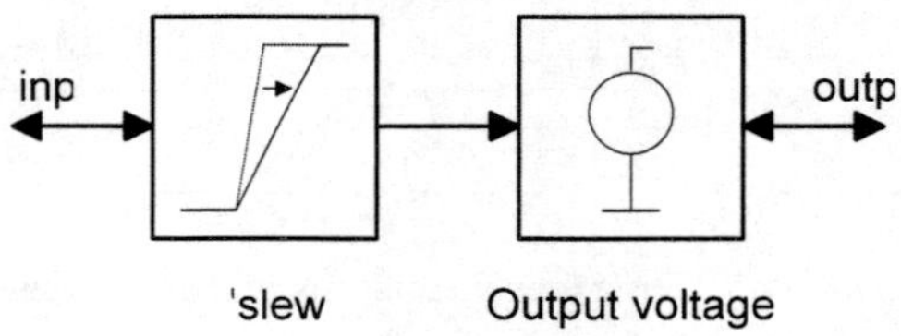

Figure 7-39. Block diagram of an ideal peak detector

The main characteristics of the peak detector model are as follows.

- The *'slew*-Attribute is applied to the input signal with an infinite slew rate for the rising edge (REAL'HIGH) and an almost zero slope for the falling edge (-10^{-38})

Additional characteristics, that were not modeled here, include input impedance, and exponential decay.

Model interface

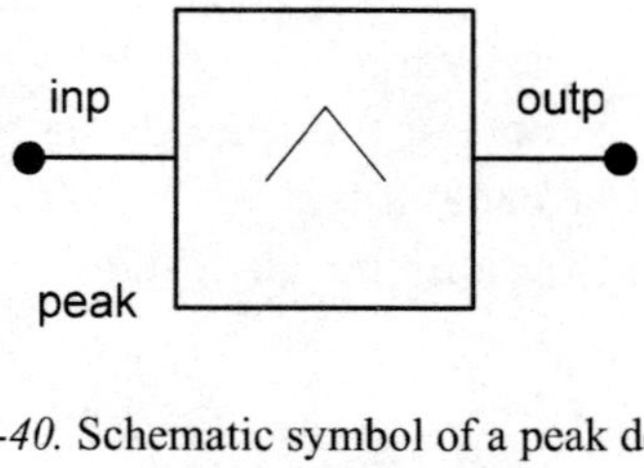

Figure 7-40. Schematic symbol of a peak detector

Table 7-32. Model ports

Name	Type	Description
INP	ELECTRICAL	Input pin
OUTP	ELECTRICAL	Output pin

There are no parameters for the peak detector model.

For a model implementation and an example test-bench see Section 6.3.5. The complete model is included on the CD-ROM that is provided with this book.

7.4.2 Frequency measurement unit

Functional description

The frequency measurement unit determines the timing distance between two rising edges for a given threshold value. The reciprocal value is output as frequency. Figure 7-41 shows the block diagram of the frequency measurement unit.

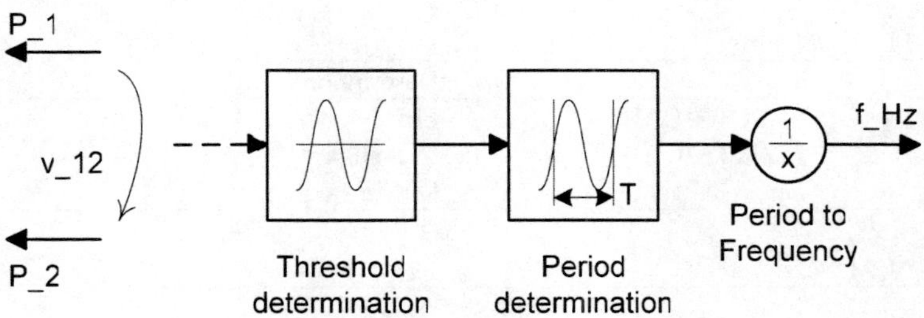

Figure 7-41. Block diagram of the frequency measurement unit

The main characteristics of the VCO model are as follows.

- The voltage across P_1 and P_2 is measured without influencing the connected network, that is no internal resistance is modeled.
- The event, when a threshold is crossed, is determined using the *'above* attribute, which leads to a time discrete signal. In this realization the threshold value is fixed and specified by a parameter of the model.
- The time discrete signal is handled in a digital process. When a rising edge is detected, the timing difference and the reciprocal value are computed.
- The resulting signal is converted back into a time continuous real quantity using the *break* statement.

- For noisy signals at the input, errors may occur due to many threshold crossings at a single positive edge.
- For the first period, the frequency output is set to zero.

Model interface

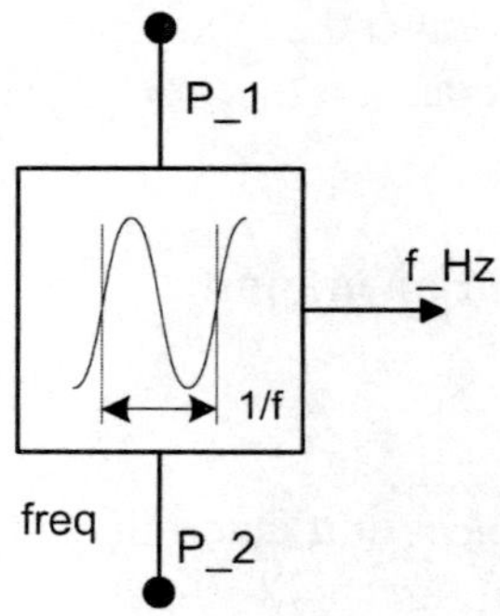

Figure 7-42. Schematic symbol of the frequency measurement block

Table 7-33. Model ports

Name	Type	Description
P_1	ELECTRICAL	First pin
P_2	ELECTRICAL	Second pin
F_HZ	out REAL	Output frequency

Table 7-34. Model parameters

Name	Unit	Default value	Description
THRESHOLD	V	0.0	Threshold value for period measurement

Model implementation

The central element of the model is the digital process, where edge detection and timing difference calculation, including initialization and reciprocal generation, take place.

```
-- digital process
process (S_12) is
  begin
    if S_12 then
      T_OLD<=NOW;
      if NOW>T_OLD and T_OLD>0.0 then
        S_FREQ<=1.0/(NOW-T_OLD);
      else
        S_FREQ<=0.0;
      end if;
    end if;
end process;
```

The complete model is included on the CD-ROM that is provided with this book.

Simulation example

For the simulation example a VCO is controlled by a sinusoidal waveform. The VCO has a free running frequency of 1.5 kHz and a sensitivity of 100 kHz/V. The controlling sine wave has an amplitude of 5 mV, which produces a frequency variation at the output of the VCO of +/-500 Hz.

The frequency measurement unit is instantiated without any parameter values, therefore the default threshold value of 0.0 applies.

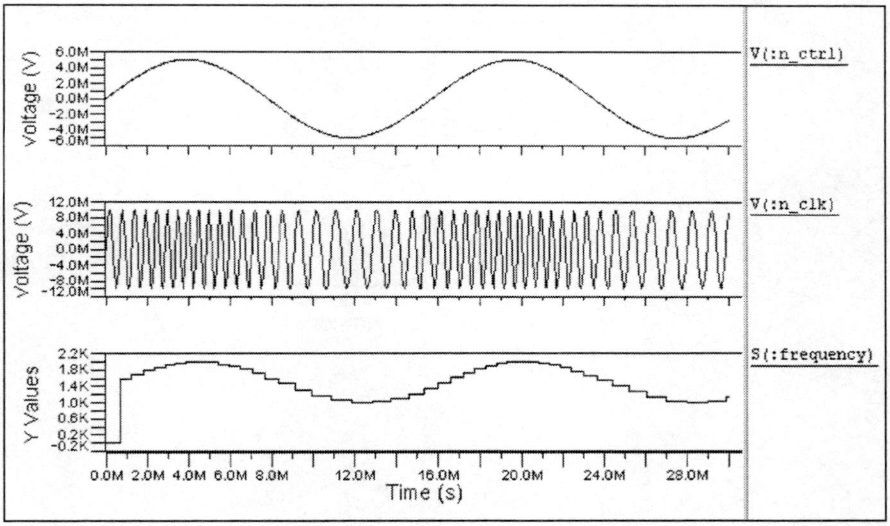

Figure 7-43. Simulation of the frequency measurement at the output of a VCO

Figure 7-43 shows the controlling sine wave (*n_ctrl*) at the input of the VCO, the frequency modulated sine wave (*n_clk*) at the output of the VCO, and the output of the frequency measurement unit (*frequency*). The quantization of that signal is due to the frequency only being determined at zero crossings of the observed signal. This also leads to a time delay between the controlling voltage and the resulting frequency measurement signal. Both effects are shown here to demonstrate the model behavior, and they are less severe at higher frequencies of the input signal.

For the first period, where the frequency is not known yet, the value of the frequency measurement signal is set to zero.

7.4.3 Power meter

Functional description

The power meter can be used as a post processing element in RF simulations. The average power of a periodic signal is computed and provided on a logarithmic scale. There are two architectures of this model for different applications: a feedthrough power meter and a terminating power meter.

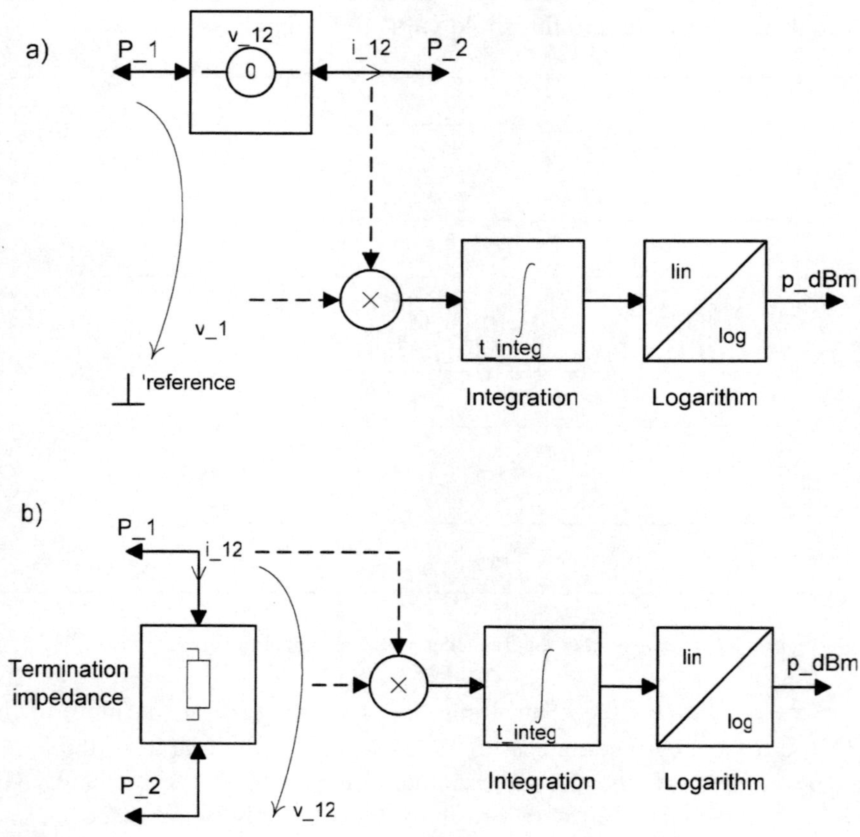

Figure 7-44. Block diagrams of power meter architectures: a) feedthrough, b) terminating

The main characteristics of the power meter model are as follows.

- Determination of the instantaneous power
  architecture *feedthrough*: $p(t) = v_1(t) \cdot i_{12}(t)$
  architecture *terminating*: $p(t) = v_{12}(t) \cdot i_{12}(t)$

- Integration over specified period

$$\overline{p}_i = \frac{1}{t\_integ} \int_{t_i}^{t_i + t\_integ} p(t)\, dt$$

Note: An input signal is assumed which is periodic with a period of t_integ, but is not necessarily sinusoidal.

- Logarithmic output as a real quantity

$$p\_dBm = 10 \cdot \log_{10}\left(\frac{\overline{p}_i}{1\text{mW}}\right), \text{ with } \overline{p}_i > 0$$

Note: The logarithmic measure can only be computed from positive power values. Therefore the model is restricted to measure the power of consumers, where the integrated power over time is positive.
For zero and negative values of power the logarithmic measure is set to -300 dBm, which corresponds to 10^{-33} W.

- Current measurement source for the *feedthrough* architecture with value $v_{12} = 0$

- Termination impedance for the *terminating* architecture, modeled as ohmic resistance: $r_{term} = 50\Omega$

Model interface

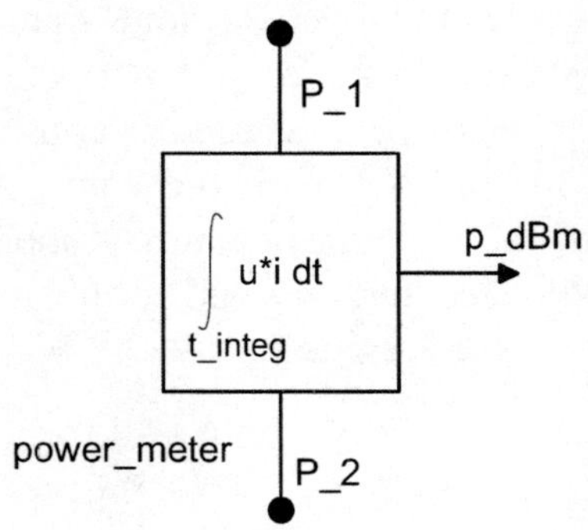

Figure 7-45. Schematic symbol of power meter

Table 7-35. Model ports

Name	Type	Description
P_1	ELECTRICAL	First measurement pin
P_2	ELECTRICAL	Second measurement pin
P_DBM	out REAL	Output power

Table 7-36. Model parameters

Name	Unit	Default value	Description
T_INTEG	s	1.0e-03	Integration period
R_TERM	Ohm	50.0	Termination resistance (feedthrough only)

Model implementation

There is little difference between both architectures. Architecture *feedthrough* uses a source with zero voltage and can therefore be used as a connection between matched blocks.

```
...
  V_12 == 0.0;
  P_INT'DOT == V_1 * I_12;
...
```

In contrast, architecture *terminating* uses a termination resistor and is therefore suitable at the end of an analog signal processing chain where matching is required.

```
...
  V_12 == I_12 * R_TERM;
  P_INT'DOT == V_12 * I_12;
...
```

The complete models of both architectures are included on the CD-ROM that is provided with this book.

Simulation example

For the simulation example a sinusoidal signal of 1 kHz is used as input, whose amplitude is modulated by a second signal of 1 Hz. Since both frequencies are so distanced from each other, the amplitude during one period of the first frequency can be regarded as constant.

This signal is fed to an LNA with 6 dB power gain and a 3rd order intercept point of 23 dBm. The signal power is monitored at the input and output of the LNA using feedthrough and terminating architectures of the power meter, respectively (see Figure 7-46).

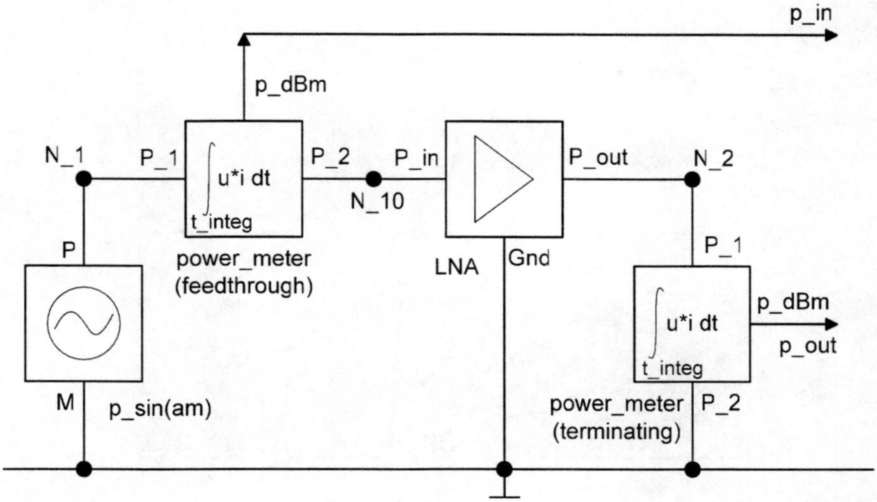

Figure 7-46. Schematic of the simulation example

The models were instantiated with the following parameters.

```
SRC1: entity P_SIN(AM)
        generic map (PA_DBM => -40.0,
                     FREQ   => 1.0e03,
                     FREQM  => 1.0,
                     MDI    => 10.0)
        port map  (P => N_1,
                   M => ELECTRICAL_REF);

LNA1: entity LNA(RF)
        generic map (GP_DB   => 6.0,
                     IP3_DBM => -23.0)
        port map  (P_IN => N_10,
                   P_OUT => N_2,
                   VDD => N_VDD,
                   GND => ELECTRICAL_REF);

PWRi: entity POWER_METER(FEEDTHROUGH)
        generic map (T_INTEG => 1.0e-03)
        port map  (P_1 => N_1,
                   P_2 => N_10,
                   P_DBM => P_IN);

PWRo: entity POWER_METER(TERMINATING)
        generic map (T_INTEG => 1.0e-03)
        port map  (P_1 => N_2,
                   P_2 => ELECTRICAL_REF,
                   P_DBM => P_OUT);
```

The complete test bench is included on the CD-ROM that is provided with this book.

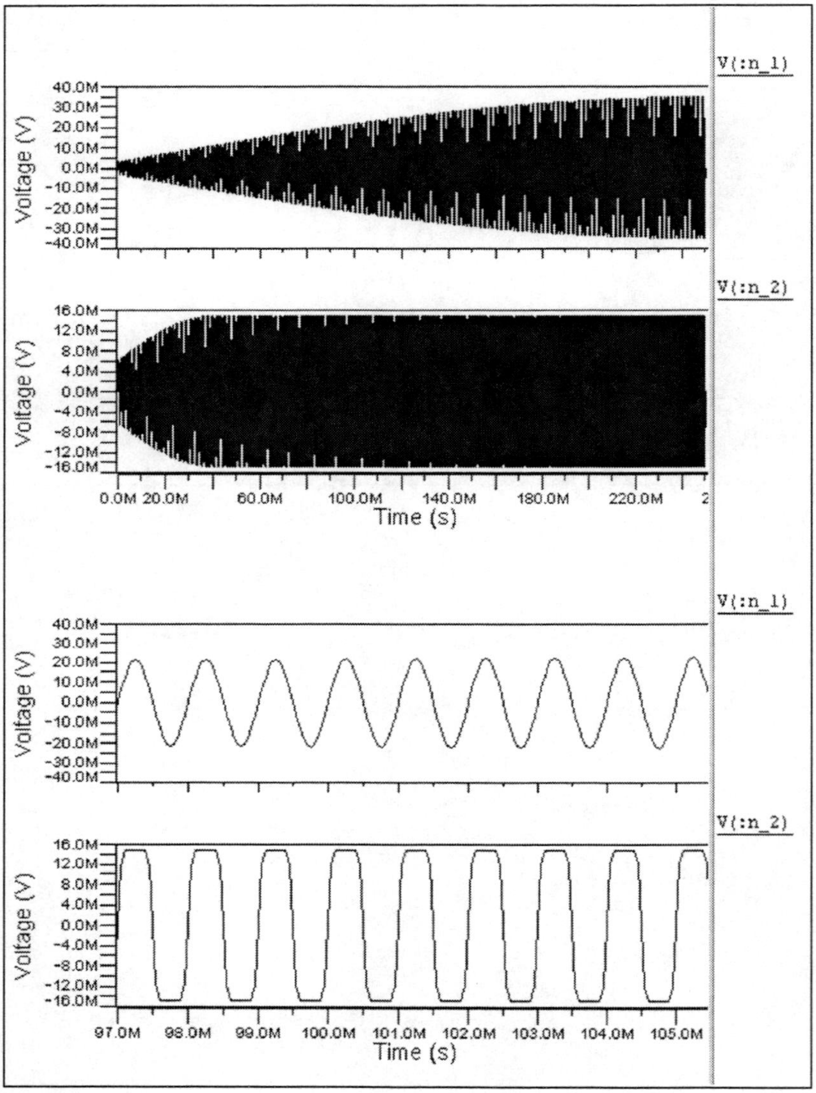

Figure 7-47. Simulation of the nonlinear characteristic of the LNA

As shown in Figure 7-47 the input signal, which should be doubled in magnitude (power gain of 6 dB), is distorted while passing the LNA. Since the amplitude of the sine wave varies slowly, it is nearly constant during a single period, as displayed in the enlarged diagram. The power of the input and output signal is measured using the power meter. The resulting transfer characteristic is displayed in Figure 7-48.

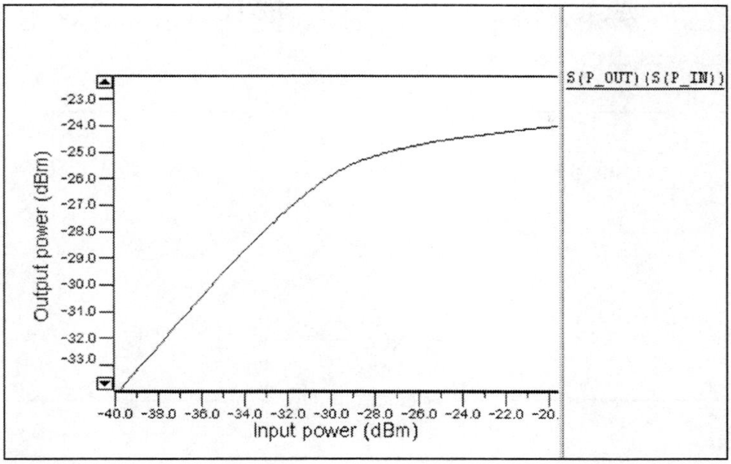

Figure 7-48. Transfer characteristic of the LNA computed with the power meter

This diagram enables improved verification of the power level behavior. In the linear region the gain is approximately 6 dB, which we expect from the specification. If the input power is increased, the output is distorted an increasing amount by the nonlinearity of the LNA. At nearly -33 dBm the gain drops to 5 dB, which is therefore said to be the 1 dB compression point.

From theory (see [Kun02]) we know that the 1 dB compression point is 9.6 dB below the IIP3 in systems where the compression point is largely determined by the 3rd order distortion. In this simulation example we specified an IIP3 for the LNA of -23 dBm and measured a 1dB compression point of nearly -33 dBm, which meets all our expectations.

7.5 Block Level Example of a Linear PLL

In this section an example is reused and extended that was the introductory example of the VHDL-AMS chapter in Section 5.3 – the linear phase-locked loop (PLL). After the previous sections introduced a library of basic RF building blocks, the PLL will be now build up from these blocks by instantiating and parameterizing them. Some background information is initially provided to give a starting point for the following detailed analysis. Then the behavior of the assembled block level example can be verified by means of simulation and mathematical computations.

Background

The phase-locked loop is a circuit where the output signal tracks the input signal with respect to phase and frequency. A PLL is used in many applications specifically as a frequency synthesizer and frequency divider.

The name *phase-locked loop* indicates the structure, which is a closed control loop as depicted in Figure 7-49.

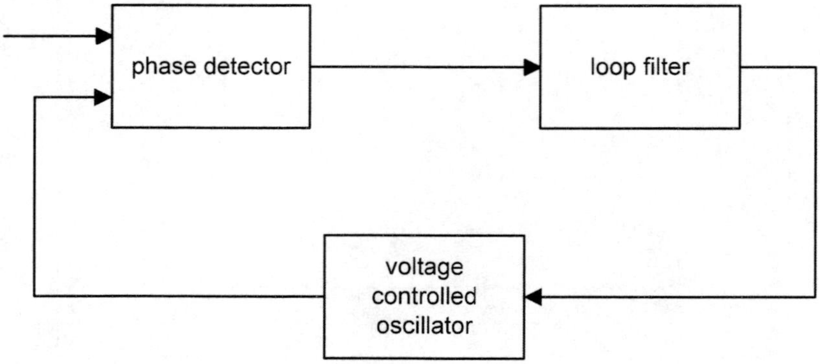

Figure 7-49. Block diagram of a phase-locked loop

The phase of the signal is compared to the phase of a reference input signal. The phase detector converts the phase difference into a proportional voltage, which is then followed by a loop filter to reject unwanted frequencies. The filtered signal is a measure for the frequency of the input signal. To close the loop, the loop filter output is fed to a voltage controlled oscillator (VCO), which generates a sine wave with proportional frequency. Now the loop starts again with a phase comparison in the phase detector. If the input frequency increases the lowpass filtered output signal rises and in turn produces a higher frequency in the VCO until the frequency of both the reference and the internal signal are equal.

Analyzing the PLL is rather complex, especially its dynamic behavior. For a detailed study of static and dynamic parameters like lock range, pull-out range, pull-in range, and hold-in range see [Bes03].

We want to keep the analysis simple and consider only a PLL which is already locked and remains locked. In this case the behavior can be treated as linear and the transfer function of the whole system can be obtained by combining the individual transfer characteristics of the building blocks. Furthermore, we restrict the calculation to a linear PLL (LPLL). Other types are known, such as digital PLL (DPLL), all-digital PLL (ADPLL), and software PLL (SPLL). For all types, the basic working principles remain the same.

Consider two sinusoidal signals, one a cosine wave from the reference source

$$u_1(t) = a_1 \cos(\omega_1 t + \varphi_1(t))$$

and the other a sine wave from the VCO

$$u_2(t) = a_2 \sin(\omega_2 t + \varphi_2(t)) .$$

Both are fed into a mixer, which is the phase detector for a linear PLL and in the simplest case is essentially a multiplier.

$$u_1(t) \cdot u_2(t) = a_1 a_2 \cos(\omega_1 t + \varphi_1(t)) \cdot \sin(\omega_2 t + \varphi_2(t))$$

In the locked state we have equal frequencies $\omega_1 = \omega_2 = \omega$, but maybe a phase error $\varphi_e(t) = \varphi_2(t) - \varphi_1(t)$ is present. With the theorem for trigonometric functions

$$\cos x \sin y = \tfrac{1}{2}\left(\sin(x+y) - \sin(x-y)\right)$$

we yield at the output of the mixer

$$\begin{aligned} u_1(t)u_2(t) &= \frac{a_1 a_2}{2}\left(\sin\left(2\omega t + \varphi_1(t) + \varphi_2(t)\right) - \sin\left(\varphi_1(t) - \varphi_2(t)\right)\right) \\ &= \frac{a_1 a_2}{2}\sin\left(\varphi_2(t) - \varphi_1(t)\right) + \frac{a_1 a_2}{2}\sin\left(2\omega t + \varphi_1(t) + \varphi_2(t)\right) \end{aligned}$$

If we suppress the term at the double frequency by filtering, and substitute for small phase errors $\sin x \approx x$, we have

$$\begin{aligned} u_d(t) &= u_1(t)u_2(t) \\ &= \frac{a_1 a_2}{2}\sin\left(\varphi_2(t) - \varphi_1(t)\right) \\ &\approx \frac{a_1 a_2}{2}\left(\varphi_2(t) - \varphi_1(t)\right) \end{aligned}$$

In the Laplace domain with $\Phi(s) = \mathfrak{L}\{\varphi_e(t)\} = \mathfrak{L}\{\varphi_2(t) - \varphi_1(t)\}$ and $K_d = a_1 a_2 / 2$ the *mixer characteristic* can be described as

$$U_d(s) \approx K_d \Phi_e(\mathrm{s})$$

For the *filter characteristic* we choose a first order lowpass filter

$$F(s) = \frac{U_f(s)}{U_d(s)} = \frac{A_0}{1 + s\tau_g} \text{ with } \tau_g = \frac{1}{2\pi f_g}$$

where A_0 is the filter gain and τ_g is the time constant of the first order pole.

The *VCO* is defined by the equation

$$u_2(t) = a_2 \sin(\omega_2 t + \varphi_2(t))$$

where the phase is controlled by the input voltage via

$$\dot{\varphi}_2(t) = K_f u_f(t)$$

Thus, the VCO behaves like an integrator if we consider the phase of the VCO signal $\varphi_2(t)$ with respect to the control voltage $u_f(t)$. In the Laplace domain the *VCO characteristic* is therefore

$$\Phi_2(s) = \frac{K_f}{s} U_f(s)$$

The theoretical system behavior of the linearized PLL model is summarized in Figure 7-50, where the block names are replaced by their transfer functions in the Laplace domain.

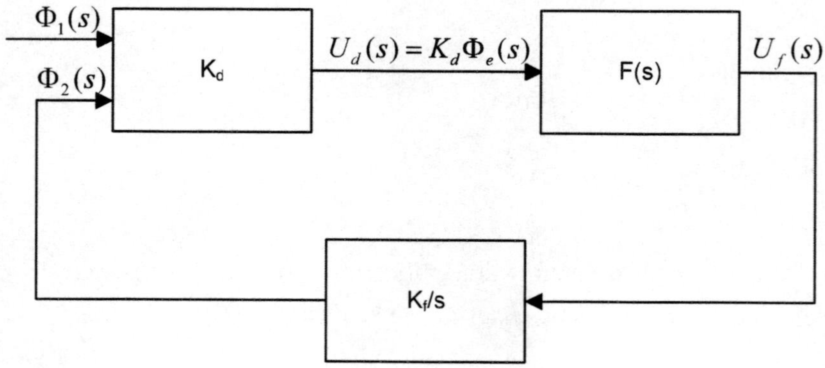

Figure 7-50. Linearized PLL model in the Laplace domain

The overall characteristic of the PLL can now be obtained using control theory, which says that the transfer function of a closed loop is the transfer function of the open loop divided by this function plus one. The phase-transfer function is therefore given by

$$\begin{aligned} H(s) &= \frac{\Phi_2(s)}{\Phi_1(s)} \\ &= \frac{F(s)K_d K_f / s}{1 + F(s)K_d K_f / s} \\ &= \frac{A_0 K_d K_f}{A_0 K_d K_f + s(1 + s\tau_g)} \end{aligned}$$

From the denominator, which is in the normalized form written as $s^2 + 2\zeta\omega_n s + \omega_n^2$, we obtain an estimate of the natural frequency ω_n and the damping factor ζ of the PLL

$$\omega_n = \sqrt{A_0 K_d K_f / \tau_g}$$

$$\zeta = \frac{1}{2\sqrt{A_0 K_d K_f \tau_g}}$$

These parameters are important to evaluate the performance and the stability of the circuit.

Objective

- A phase-locked loop will be constructed according to the block diagram in Figure 7-49 using the following list of parameters.

Table 7-37. PLL block parameters

Block	Parameter	Value
Mixer	Power gain	12.0 dB
	IP3	-10 dBm
Filter	Gain	0 dB
	Corner frequency	100 kHz
	Filter grade	1
VCO	Amplitude	-30 dBm
	Free running frequency	1 MHz
	Sensitivity	1 Hz/V

- A reference signal is applied to the input of the PLL using a piecewise linear voltage source and a VCO with a cosine output signal and the following list of parameters.

Table 7-38. PLL test-bench parameters

Block	Parameter	Value
Piecewise linear source	Amplitude 0 s (initial)	0 μV
	Amplitude 100 μs	50 μV
	Amplitude 200 μs	100 μV
VCO	Amplitude	-30 dBm
	Free running frequency	1 MHz
	Sensitivity	1 GHz/V

- The PLL is simulated over 300 μs with particular attention to
  - input control signal
  - input reference clock signal
  - output clock signal (output of the PLL internal VCO)
  - lowpass filtered signal
- With the specified set of parameters the control loop of the PLL should perform a damped oscillation when the input reference signal is abruptly changed. This oscillation can be observed at the output of the filter and has a period T_r. Figure 7-51shows the period of the damped oscillation.

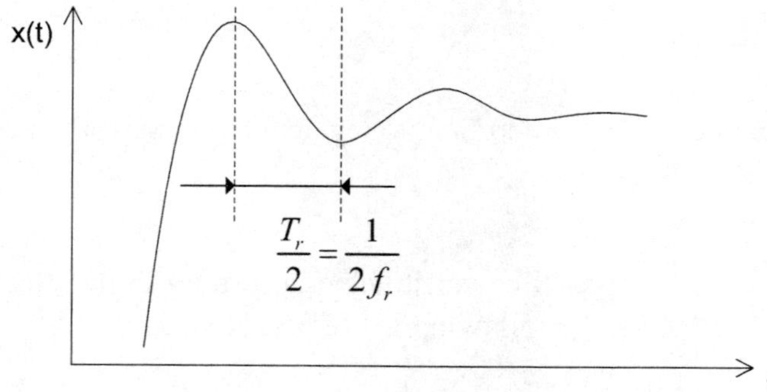

Figure 7-51. Period measurement of the damped oscillation

- The PLL model can then be verified by mathematical analysis. The theoretical period of the damped oscillation can be computed from the model parameters of the PLL using the relationship for the angular frequency $\omega_r^2 = \omega_n^2(1-\zeta^2)$, where the natural frequency ω_n and the damping factor ζ are specified above. The theoretical period will be compared to the simulation result.
  Note: Since we use a first order filter here for simplicity, there will still be some oscillation on the filtered signal. Nevertheless an estimation of the period should be possible, which is sufficient to verify the fundamental conformity.

Solution

- The complete models of the PLL and the test-bench for the example solution are included on the CD-ROM that is provided with this book.
- Simulation results for the control voltage at node *N_ctrl* and for the output of the lowpass filter *N_tp* are shown in Figure 7-52.

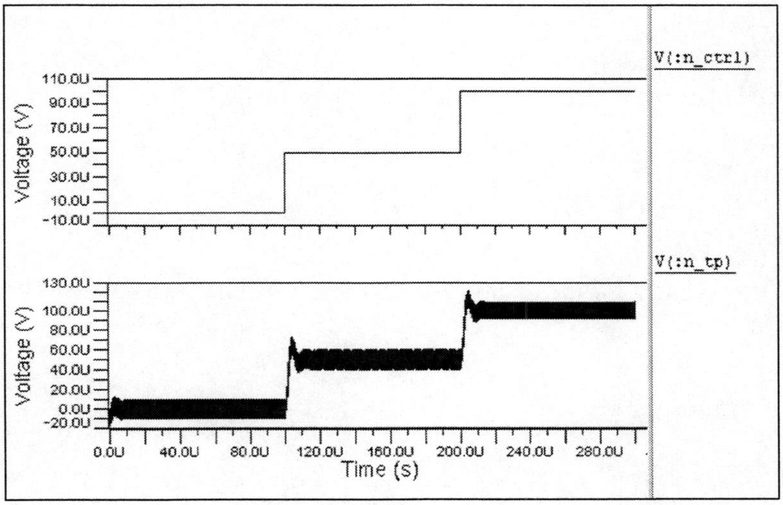

Figure 7-52. Results of the PLL simulation

- The period of the damped oscillation can be estimated to $T_f/2 \approx 3.8\,\mu\text{s}$ from the simulation results in Figure 7-53.

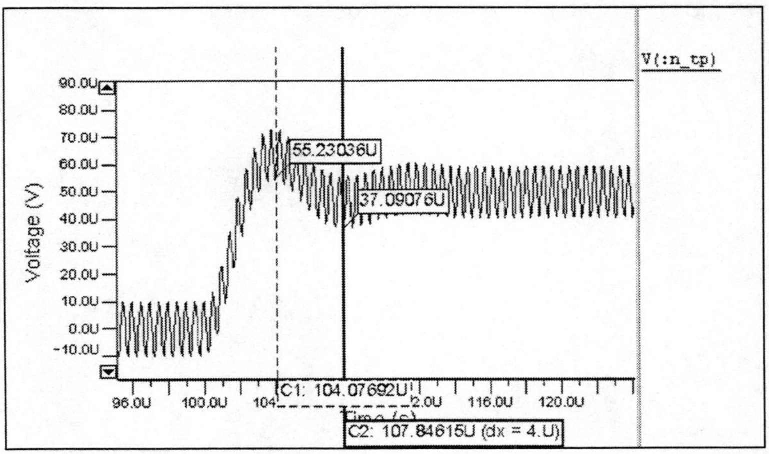

Figure 7-53. Period measurement in the simulation results

- For the exact computation of the frequency we need the individual transfer factors of the blocks

$$A_0 = 1$$
$$K_f = 2\pi \cdot 10^9 \text{ Hz/V}$$
$$K_d = gain_{mixer} \frac{a_1 a_2}{2}$$
$$= 10^{gp\_dB/10} \cdot \frac{1}{2} \cdot \left(\sqrt{10^{ampl\_dB/10}} \cdot 2 \cdot R \right)^2$$
$$\approx 0.2\text{mV}$$
$$\tau_g = 1/(2\pi \cdot 10^5 \text{Hz})$$

- Consequently for the damped oscillation frequency we get

$$\omega_n^2 = A_0 K_d K_f / \tau_g$$
$$\zeta^2 = 1/(4 A_0 K_d K_f \tau_g)$$
$$\omega_r^2 = \omega_n^2 (1 - \zeta^2)$$
$$= \frac{A_0 K_d K_f}{\tau_g} - \frac{1}{4\tau_g^2}$$
$$f_r = \omega_r / 2\pi \approx 132\text{kHz}$$

This yields a theoretical value for the half period of $\overline{T}_f / 2 = 3.8\mu\text{s}$, which matches the value obtained from the simulation results.

Chapter 8

MACROMODELING IN VHDL-AMS

8.1 Introduction

The term macromodeling has been used for different meanings in the past. Therefore the term we use here is first clarified together with an overview of the modeling method.

Afterwards, a set of building blocks that can be used for macromodeling is presented. We use the VHDL-AMS hardware description language here to describe the behavior of the blocks.

Using these building blocks a simple but extendible macromodel is constructed for the operational amplifier (OpAmp), which is a very important system component.

8.2 General Methodology

Macromodeling is a well known and frequently used modeling method for many years [BCP74], [CaS91]. Historically the term is used for structural modeling for SPICE-like network simulators [CoC92]. It is assumed that the original circuit can be subdivided into smaller blocks (macros) which are describable independently of each other, that is, they are only weakly coupled. Each block of the circuit contains a number of active and passive elements. During the modeling step the block is replaced by a network of ideal controlled sources and other ideal basic elements. Thus the resulting macromodel consists of blocks which are individually modeled by idealized network elements to represent a particular functionality of the overall circuit.

Presently macromodeling is mostly used in a wider sense. The circuit is divided again into smaller weakly coupled subsystems. However, these blocks or macros can be modeled by many different means of description, including:

- Substitution network with idealized network elements
- Behavioral description with suitable HDL
- Hierarchical composition of behavioral and structural descriptions
- Coupling to a specialized simulator

A very common method of subdividing a circuit into blocks is shown in Figure 8-1.

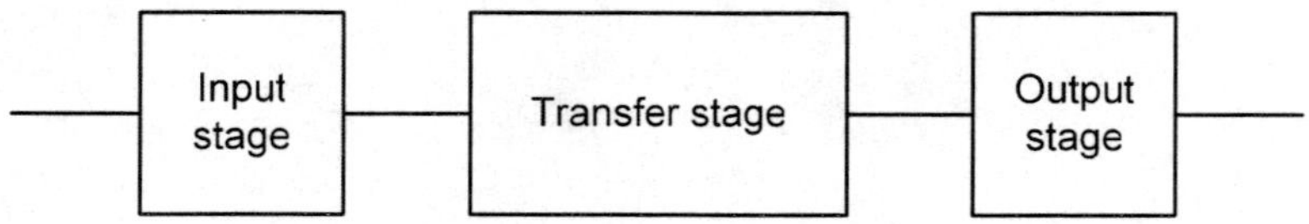

Figure 8-1. General structure of a macromodel

The input and output stages are modeled very precisely using, for instance, the first and last transistor stages from the circuit, respectively. Between both stages the functional characteristic is modeled by a transfer stage. This subdivision has a number of advantages:

- Since input and output stages use the same transistors at the border as the circuit, the model "looks like" the circuit from the outside world of the surrounding circuitry.
- This implies that the circuit and model can be exchanged with each other, which is called pin-compatibility, because both are compatible with respect to their pins.
- In the transfer stage the functionality can be modeled in a very abstract way without too many details, since it is not directly connected to the outside world.
- This significantly saves computing time and therefore even allows simulation of very complex circuits together with their overall applications.

The separation of the functionality from the pin behavior is an important aspect for choosing the appropriate means of description. For input and output stages electrical terminals with conservative behavior must be used, whereas the transfer blocks can be described with non-conservative signals. This allows to apply signal flow models, or even data driven descriptions, that are especially well suited for functional blocks.

The methodology described proves especially useful for digital circuits in an analog environment. In this case the input and output stages remain analog, whereas the transfer stage is modeled using abstract digital means of description (for example logic gates, register transfers). Between the stages a-to-d and d-to-a converters must be inserted. Usually it is sufficient to use simple conversion characteristics for the converters (see Figure 8-2).

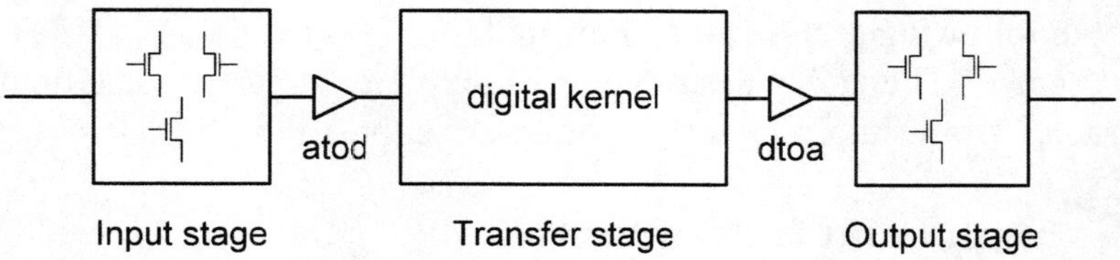

Figure 8-2. Structure of a mixed-signal macromodel

In a chain of digital blocks the d-to-a converters at the output and the a-to-d converters at the input of the next stage can be omitted, provided that the internal signals do not need to be monitored.

The macromodels described require a true mixed-signal simulation. Behavioral descriptions of these macromodels are best supported by mixed-signal HDLs, such as VHDL-AMS. They allow seamless integration of analog and digital together with conservative and non-conservative parts into one behavioral model.

To further generalize the macromodeling principle we can extend the macromodel structure as exemplified in Figure 8-3.

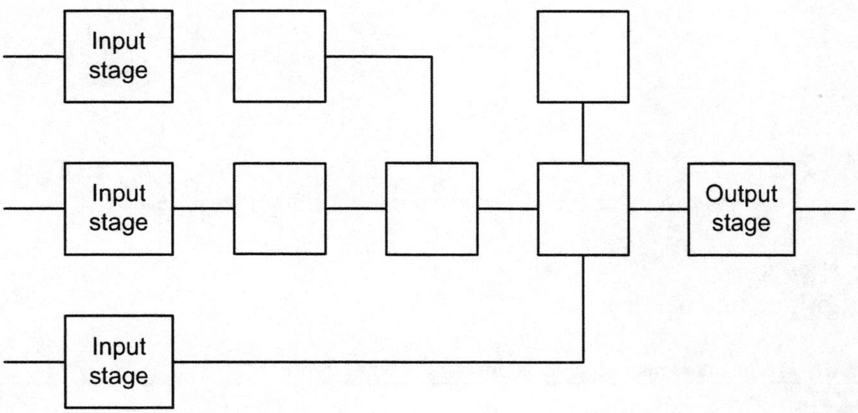

Figure 8-3. Extended structure of a macromodel

Between the input and output stages several blocks are placed, each representing a part of the overall functionality (for example biasing, linear gain, nonlinear distortion, frequency characteristic, current/voltage limitations, power supply dependency).

The key point is the assumption that certain functions of the circuit can be individually modeled independently of each other and then superimposed to form the resulting macromodel.

In the following section typical examples of input and output (I/O) stages are presented. Thereafter a simple example of an OpAmp macromodel is constructed from simplified stages for input, output, and transfer.

8.3 Input and Output Stages

The macromodeling methodology presented in the previous section is a common way to describe integrated circuits for system simulation. For input and output stages typical building blocks can be defined that are usable for many applications.

This section provides building blocks by means of a schematic of the stage followed by a list of properties and a table detailing the VHDL-AMS implementation.

8.3.1 Input stages

Ideal differential input

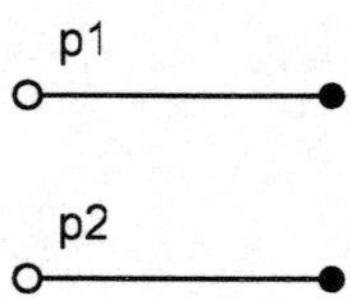

Figure 8-4. Network schematic of an ideal differential input stage

Properties

- Infinite internal resistance

Table 8-1. Implementation of an ideal differential input stage

VHDL-AMS notation
`quantity V across P1 to P2;`

Real differential input

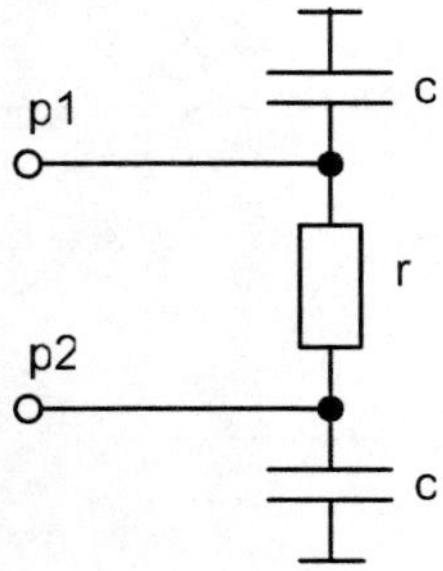

Figure 8-5. Network schematic for a real differential input stage

Properties

- Finite internal resistance
- Differential input resistance
- Input capacitances

Table 8-2. Implementation of a real differential input stage

VHDL-AMS notation

`quantity V1 across I1 through P1;`
`quantity V2 across I2 through P2;`
`quantity V12 across I12 through P1 to P2;`
`I1 == C * V1'DOT;`
`I2 == C * V2'DOT;`
`V12 == R * I12;`

Differential input with offset

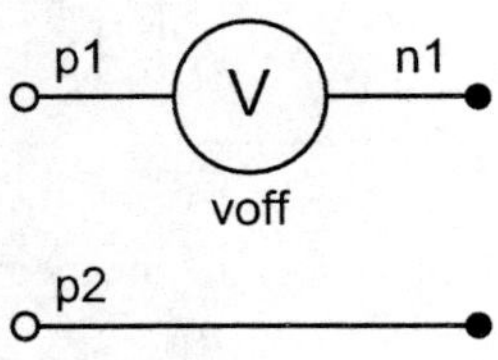

Figure 8-6. Network schematic for a differential input stage with offset

Properties

- Offset voltage
- Infinite internal resistance

Table 8-3. Implementation of a differential input with offset

VHDL-AMS notation

`quantity V11 across I11 through P1 to N1;`
`quantity V12 across N1 to P2;`
`V11 == VOFF;`

Termination resistor to ground

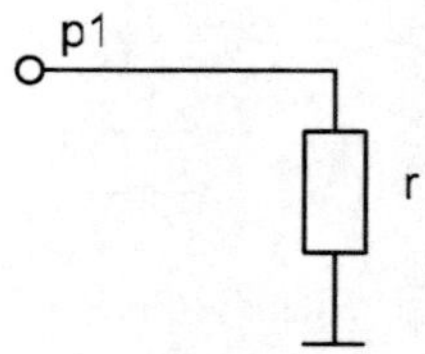

Figure 8-7. Network schematic for a termination resistor to ground

Properties
- Single ended pin

Table 8-4. Implementation of a termination resistor to ground

VHDL-AMS notation
`quantity V1 across I1 through P1;` `V1 == R * I1;`

Termination resistor to supply voltage

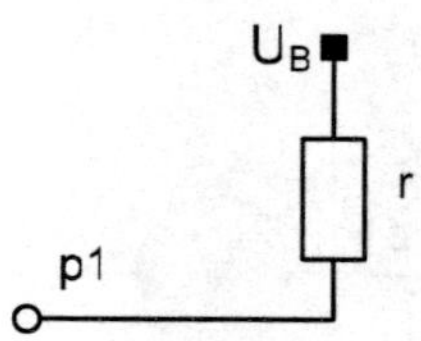

Figure 8-8. Network schematic for a termination resistor to supply voltage

Properties
- Single ended pin

Table 8-5. Implementation of a termination resistor to supply voltage

VHDL-AMS notation
`quantity V1 across I1 through P1 to NUB;` `V1 == R * I1;`

8.3.2 Output stages

Ideal output

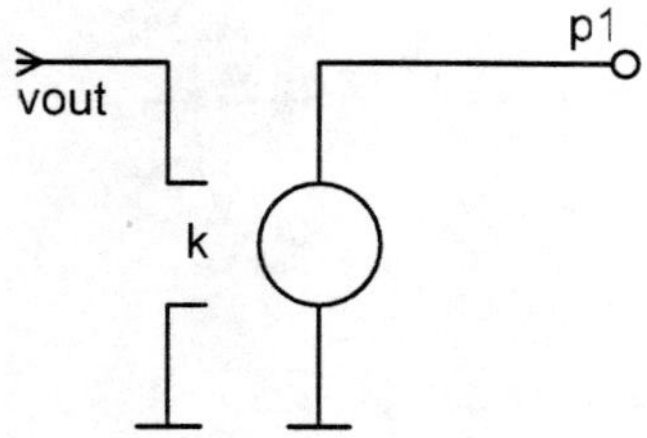

Figure 8-9. Network schematic for an ideal output

Properties

- Controlled source
- Zero internal resistance

Table 8-6. Implementation of an ideal output

VHDL-AMS notation
quantity V1 **across** I1 **through** P1; V1 == K * VOUT;

Real output

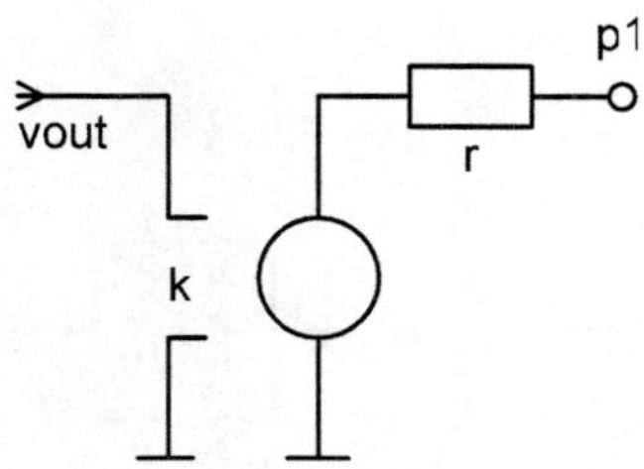

Figure 8-10. Network schematic for a real output

Properties

- Controlled source
- Internal resistance

Table 8-7. Implementation of a real output

VHDL-AMS notation
quantity V1 **across** I1 **through** P1; V1 == K * VOUT + R * I1;

Voltage limitation, structural

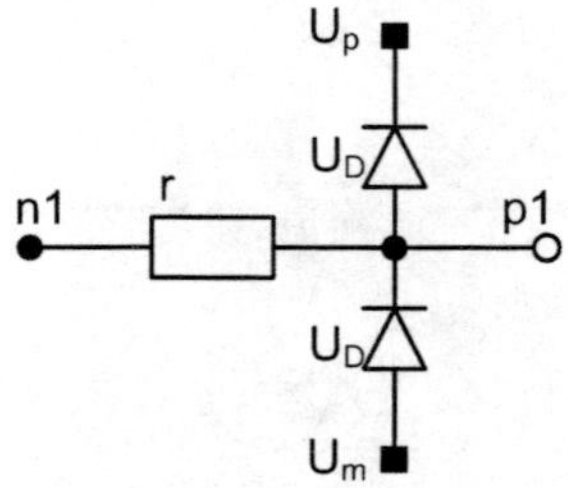

Figure 8-11. Network schematic for a structural voltage limitation stage

Properties

- Limitation to $U_m - U_D < U < U_p + U_D$ with smooth transition
- No decoupling
- Very high currents through the diodes may occur in case of limitation

Table 8-8. Implementation of a structural voltage limitation

VHDL-AMS notation

```
quantity V1 across I1 through N1 to P1;
quantity VP across IP through p1 to NUP;
quantity VM across IM through NUM to P1;
  V1 == R * I1;
  IP == I0 * EXP(VP / UT - 1.0);
  IM == I0 * EXP(VM / UT - 1.0);
```

Voltage limitation, behavioral

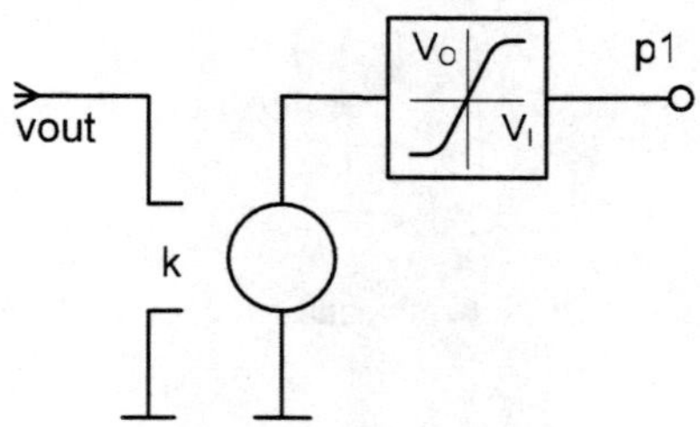

Figure 8-12. Network schematic for a behavioral voltage limitation stage

Properties

- Controlled source
- Different limiting functions possible, for example piece-wise linear, polynomial, logarithmic, tanh

Table 8-9. Implementation of a behavioral voltage limitation

VHDL-AMS notation
`quantity V1 across I1 through P1;` `  V1 == k * VOUT - C * VOUT**3;`

Current limitation, behavioral

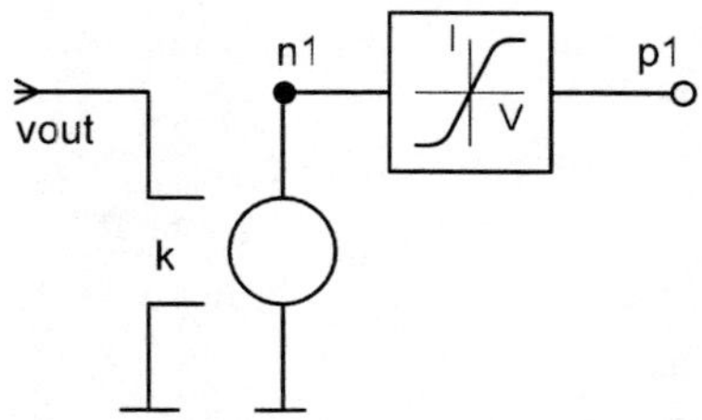

Figure 8-13. Network schematic for a behavioral current limitation stage

Properties

- Controlled source
- Internal resistance depending on current flowing through it

Table 8-10. Implementation of a behavioral current limitation

VHDL-AMS notation

`quantity V1 across I1 through N1;`
`quantity VR across IR through N1 to P1;`
`  V1 == K * VOUT;`
`  IR == IMAX * TANH(VR/ROUT/IMAX);`

The presented I/O macros in VHDL-AMS are just a few examples that typically occur in macromodels. They can be used in combination, adjusted with different functional forms, or extended to the specific needs of the designer.

The source code shown includes only the most necessary lines. Additional effort must be made for a working model in terms of entity and architecture structure, terminal declarations, and so on. However, since these lines of code are meant as macros, they should be inserted into working models.

8.4 OpAmp Macromodel

The operational amplifier is a building block that is very often used in analog circuits. It gained its name from the initial application purpose in analog computers. The OpAmp can easily be used for mathematical operations like summing, integrating, or comparing (electrical) values.

Although we use digital computers today, the OpAmp is applied everywhere in analog signal processing or signal conditioning, for example signal amplification, integration, buffering, impedance transformation, and so on. In most applications the OpAmp works within a feedback loop where the nominal amplification is not the primary focus. Instead linearity and large signal behavior have to be considered.

The ideal operational amplifier can be characterized as follows:

- Infinite open loop gain and bandwidth
- Infinite common mode rejection ratio (CMRR)
- Infinite differential and common mode input resistance
- Zero output resistance
- Negligible offset voltage and input current
- No noise and no feedback from the output to the input

Real OpAmps possess finite values for these characteristics.

When modeling OpAmps the choice of which characteristics have to be included depends on the intended purpose. Usually we differentiate between first order effects that are essential for the function of the block (for example gain) and second order effects that can sometimes be neglected (for example thermal behavior). A list of examples for both categories is provided next.

First order characteristics
- Open loop gain
- Corner frequency (3dB frequency)
- Output limitation
- Input and output impedances

Second order characteristics
- Offset voltage
- Power dissipation
- Common mode voltage and common mode range
- Common mode rejection ratio (CMRR)
- Power supply rejection ratio (PSRR)
- Slewrate
- Settling time
- Signal to noise ratio (SNR)
- Higher order poles and zeros of the transfer function

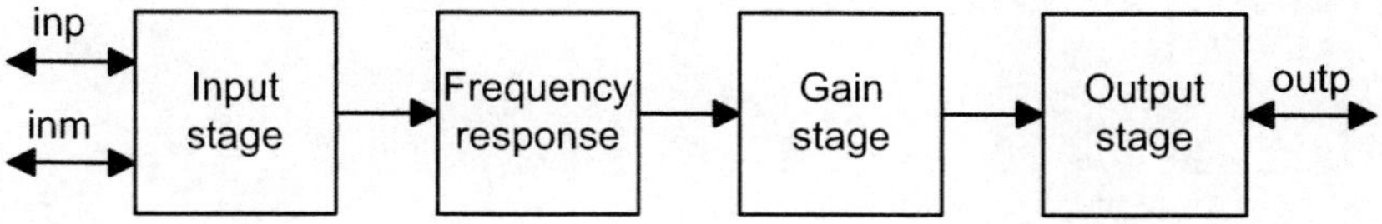

Figure 8-14. Block diagram of a simple OpAmp macromodel

Figure 8-14 shows the block diagram of a simple OpAmp macromodel, where mainly first order effects are included. Each individual block can be refined according to the required accuracy of the model.

Input and output stages may be selected from the library in the previous section, Section 8.2. They are responsible for the electrical behavior of the model to the outside world. Therefore conservative pins are used for *inp*, *inm*, and *outp*. Between the stages non-conservative signals can also be used. This prevents feedback from the output to the input, which is not always an advantage in analog modeling.

According to the macromodeling strategy more effects can now be added if required. Examples could include additional blocks for power dissipation and power supply rejection. This would require additional power supply pins, which again must be conservative since they have contact with the surrounding circuitry.

As an example the VHDL-AMS macromodel of an operational amplifier is provided below.

```
-- -------------------------------------------------------
-- Description: Behavioral model for Operational Amplifier
-- -------------------------------------------------------
library IEEE;
  use IEEE.ELECTRICAL_SYSTEMS.all;
  use IEEE.MATH_REAL.all;

entity OPAMP is
  generic (AVD0 : REAL := 106.0;    -- DC differential gain [dB]
           FP1  : REAL := 5000.0;   -- dominant pole [Hz]
           FP2  : REAL := 2.0E6;    -- pole frequency [Hz]
           ROUT : REAL := 75.0      -- output resistance [Ohm]
           );
  port (terminal INP: ELECTRICAL;   -- input plus terminal
        terminal INM: ELECTRICAL;   -- input minus terminal
        terminal OUTP: ELECTRICAL   -- output terminal
        );
end entity OPAMP;

architecture MACRO of OPAMP is

-- Input stage
  quantity V_IN across INP to INM;

-- Frequency Response
  constant NUM_2 : REAL_VECTOR := (0 => 1.0);
```

```
  constant DEN_2 : REAL_VECTOR := (1.0, 1.0/MATH_2_PI/FP2);
  quantity q_fr3 : REAL;

-- Gain stage
  constant AVD0_VAL : REAL := 10.0**(Avd0/20.0);
  constant NUM_1 : REAL_VECTOR := (0 => 1.0);
  constant DEN_1 : REAL_VECTOR := (1.0, 1.0/MATH_2_PI/FP1);
  quantity Q_SUM : REAL;
  quantity Q_FP1 : REAL;

-- Output stage
  quantity v_out across i_out through outp;

begin

-- Input stage
  I_IN == 0.0;

-- Frequency Response
  Q_FR3 == V_IN'LTF(NUM_2, DEN_2);

-- Gain stage
  Q_SUM == AVD0_VAL*Q_FR3;
  Q_FP1 == Q_SUM'LTF(NUM_1, DEN_1);

-- Output stage
  I_OUT == (V_OUT - Q_FP1)/ROUT;

end architecture MACRO;
```

In this example the input stage is modeled simply as an ideal differential input with infinite resistance and no capacitances. The following frequency stage represents a second pole and has a non-conservative real quantity as its output. This is passed to the gain stage, which is responsible for linear amplification of the signal and the dominant first pole. In the output stage we have conservative signals again and a driving voltage source with internal resistance.

This simple demonstrative OpAmp macromodel contains only the most essential functionality. It could be modified or extended by adding additional stages to include the previously mentioned second order effects.

Chapter 9

COMPLEX EXAMPLE: WLAN RECEIVER

9.1 Introduction

Mixed-signal modeling and simulation support both top-down design and bottom-up verification. A general flow from system level to transistor or layout level and back to system level was introduced in Chapter 2 and detailed in the subsequent chapters.

In this chapter the modeling methodology is applied to a practical design example from industry. The example shows in detail how modeling and simulation can support designers' tasks.

All modeling is performed with VHDL-AMS using the RF library presented in previous chapters.

To demonstrate the major aspects of RF modeling and simulation a realistic design example for an industrial RF application is provided. It serves as a complex example demonstrating how modeling and simulation can support the work of RF designers. The following aims are addressed:

- Application of the VHDL-AMS hardware description language for behavioral and hierarchical modeling of complex circuits
- Usage of an industrial design case instead of trivial examples to demonstrate the benefits of modeling and simulation in the design flow
- Seamless integration of analog and digital parts of a circuit into a simulation of the overall behavior
- Analysis of RF specific circuit level impairments on the system level performance

The complex design example is based on a system level architecture for wireless high-speed data transmission according to the wireless local area network (WLAN) standard IEEE 802.11a. For educational purposes this example has been simplified in a way that it might not conform to the standard. The example concentrates on the important parameters and neglects some effects to keep the example manageable.

First the specification of the design case is covered in detail in Section 9.2. This informal specification would be the RF part of the system specification provided to the RF design engineer. Next, as described in the top-down design flow, the system level description has to be modeled to obtain an executable specification, see Section 9.3. It can then be refined towards a circuit level model for each individual block. In the example provided, we begin with a circuit level simulation thereby simplifying the transition from system to circuit level. Using knowledge from the circuit level implementation, the behavioral models can be calibrated as it is shown in Section 9.4. The overall system behavior is then simulated with behavioral models for the blocks to analyze whether the designed system matches the one specified in terms of critical parameters. This verification is covered in Section 9.5.

9.2 Example Specification

The WLAN receiver design example consists of the RF receiver shown in Figure 9-1.

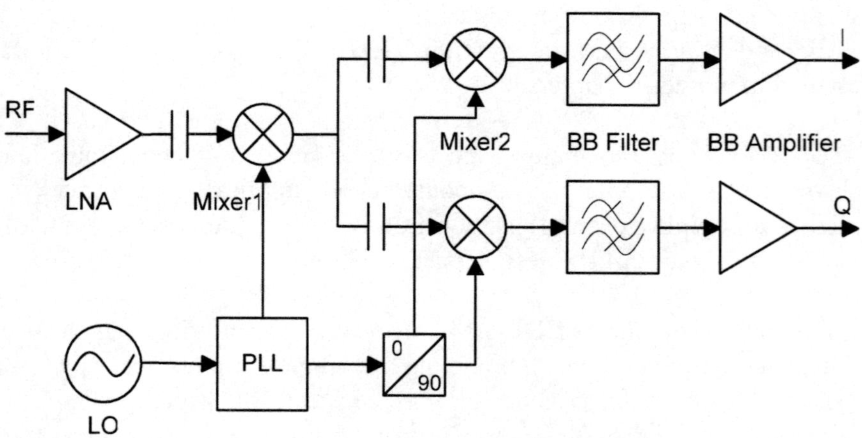

Figure 9-1. Double conversion receiver architecture

The receiver exhibits a heterodyne architecture that is frequently used in wireless and radio systems. As a special case the intermediate frequency (IF) at the output of the first mixer is chosen to be half the radio frequency, resulting in an architecture known as a *double conversion receiver.* The main components of the architecture are a low-noise amplifier (LNA), mixer (Mixer1 and Mixer2), baseband filter (BB filter), baseband amplifier (BB amplifier), and frequency synthesizer. The frequency synthesizer consists of a reference oscillator (LO), a phase-locked loop (PLL), and a 90° phase shifter.

The incoming RF signal is assumed to be a 5.2 GHz OFDM (Orthogonal Frequency Division Multiplex) signal. For simplification of this example, effects of the antenna and duplexer are neglected. After signal amplification by a low-noise amplifier (LNA) the signal is downconverted by two mixer stages, both working at the same local oscillator (LO) frequency of 2.6 GHz. The first mixer stage converts the RF input signal to half of the RF frequency, with an image frequency around zero. As there is no signal at 0 Hz, this architecture overcomes problems concerning image rejection. At the second mixer stage the RF input signal and the LO signal both have the same frequency and therefore DC-problems exist, caused by the selfmixing products. DC-offsets and flicker noise (1/f) are filtered out by highpass filtering using the capacitors between the mixer stages. In the analog baseband section channel selection is done by lowpass filtering, suppressing the adjacent and non-adjacent channels. After filtering the signal is amplified by an automatic gain controlled (AGC) amplifier. The following stages of the receiver, that is, analog-digital converter (ADC) and digital baseband, are not considered in this complex design example. Table 9-1 shows the specification of the input signal. The specification of the individual blocks is shown in Table 9-2.

Table 9-1. Input signal specification

Parameter	Unit	Channel	1st adjacent	2nd adjacent
signal power	dBm	-88 … -23 (=ch)	ch + 16	ch + 32
lower 3dB BB frequency	MHz	-8.3 … 8.3	11.7 … 28.3	31.7 … 48.3

Table 9-2. Block specification

Parameter	Unit	LNA	Mixer 1	Mixer 2	BB Filter	BB Amp
RF input	GHz	5.18 … 5.32	5.18 … 5.32	2.59 … 2.66	BB	BB
R_in	Ohm	50	50	50	50	50
input level	dBm	-88 … -23	-68 … -18	-63 … -13	-58 … -8	-58 … -8
gain	dB	5 … 20	5	5	0	12 … 62
noise figure	dB	2.5	10	10		5
CP1dB	dBm	-20				
IIP2	dBm		20	20		
IIP3	dBm	-5	5	5		

The input level of the *LNA* is between –88 dBm and –23 dBm. This input signal contains the channel signal and the signals in the adjacent (1st adjacent) or non-adjacent (2nd adjacent) channels. Depending on this input level the LNA must provide a gain between 5 dB and 20 dB. The maximum output power must be below –15 dBm to avoid overloading of the following stage.

Both *mixer stages* have the same LO frequency of 2.6 GHz applied, that is half of the carrier frequency. As a consequence, the channel signal is around 0 Hz after being downconverted two times. The RF input frequency is between 5.18 GHz and 5.32 GHz for the first mixer and between 2.59 GHz and 2.66 GHz for the second mixer. The local oscillator (LO) input frequency is between 2.59 GHz and 2.66 GHz for both mixers. Both IIP2 (input referred 2nd order intercept point, see Section 10.4.2 for details) and noise figure are important parameters for the mixers.

The *baseband filter* is a bandpass filter used for channel selection. DC offsets and adjacent and non-adjacent channels have to be rejected to a level that ensures a sufficient sensitivity for the channel signal. In OFDM systems, the center subcarrier contains no data. This allows DC offsets to be filtered out. The distance from the center subcarrier to the next data subcarrier is 156.25 kHz, resulting in a corner frequency of 150 kHz at the highpass pole.

Attenuation of the adjacent and non-adjacent channels is achieved using lowpass filters. The highest frequency of the wanted channel is around 8.3 MHz, so that the lowpass poles are positioned at a frequency of 9 MHz. A fifth-order lowpass filter is sufficient to reject the unwanted channels.

The gain of the *baseband amplifier* (AGC) is dependent on the signal level at its input. The output of the amplifier should not exceed 1 Vpp (peak-to-peak voltage), which means 4dBm assuming an output resistance of 50 Ω.

The *frequency synthesizer* generates the local oscillator signals for the RF mixer and for both I/Q mixers. Using the double conversion architecture allows the same LO frequency to be used. The orthogonality of the local oscillator signal generation for the I and Q components is a critical parameter, as a phase shift of 90° is required to guarantee a sufficient image rejection of the receiver. The same applies for the phase noise due to the low subcarrier spacing in the targeted IEEE 802.11a OFDM transmission.

The frequency synthesizer is realized by a phase-locked loop structure, including a voltage controlled oscillator (VCO), and a phase shifting block. For simplicity the frequency synthesizer is modeled with the parameters shown in Table 9-3.

Table 9-3. Specification of the frequency synthesizer

Parameter	Unit	Value (range)
tunable center frequency	GHz	2.59-2.66
phase noise (at 2 MHz)	dBc/Hz	-103
phase noise (at 20 MHz)	dBc/Hz	-121
orthogonality error	Degrees	2

9.3 Example Modeling

The top-down methodology begins with a very rough view of the transmission system as shown in Figure 9-2. A signal – which is not yet specified – has to be carried from a source to a sink by means of a limited transmission channel. These limitations require the use of a transmitter before and a receiver after the channel in order to compensate the non-ideal properties of the transmission medium. At the sending end (transmitter) the signal is coded and modulated, while at the opposite end (receiver) matched demodulators and decoders are required.

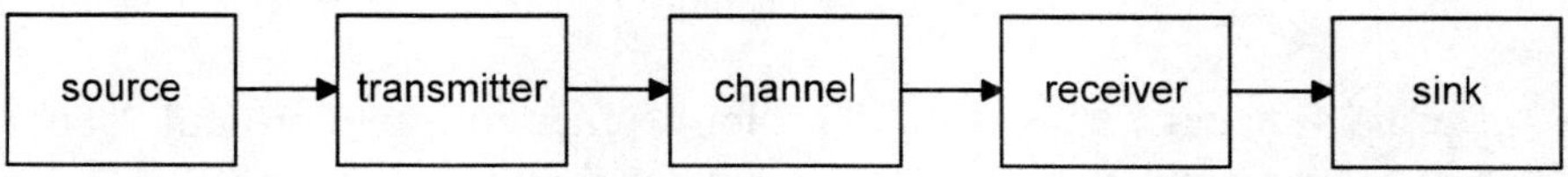

Figure 9-2. General view of a transmission system

According to the specification of the discussed example the transmission is performed in a wireless radio frequency propagation channel. A number of channel impairments may disturb the transmission, including:

- Additive white Gaussian noise (AWGN)
- Fading effects and echoes due to reflection
- Damping
- Nonlinear distortion
- Doppler effects

When refining the models we focus on the *receiver*. This part of the system usually requires the most effort to implement since timing information of the signal is not available and therefore the signal has to be reconstructed after being distorted on the channel.

The receiver is constructed with elements of the RF library introduced in previous chapters. These are behavioral models or hierarchical compositions of behavioral models written in VHDL-AMS. Figure 9-3 shows the structure of the receiver that is similar to the one specified except for the additional splitter blocks. The VHDL-AMS model implementation of the receiver can be found on the CD-ROM that is provided with this book.

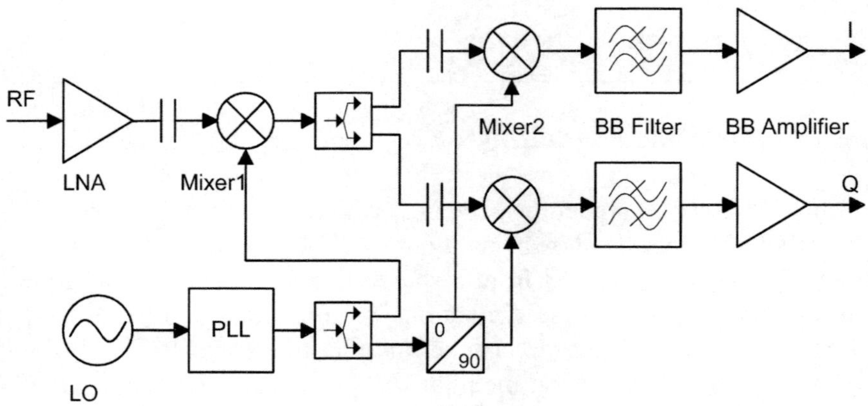

Figure 9-3. Block structure of the modeled receiver

The *LNA* model (see Section 7.3.1 for detailed information) only accepts the 3rd order intercept point as a parameter for the nonlinearity. The 1dB compression point cannot be chosen independently of IP3, since the model uses a single polynomial nonlinearity. For a detailed analysis of the same WLAN receiver example using RF specific model parameters and simulation modes see Chapter 10.

Gain and input impedance are model parameters and are set accordingly, while input frequency and level are not parameters of the model.

For *mixers* 1 and 2 an identical model is used from the library (see Section 7.3.2), which internally contains an LNA model for gain and nonlinearity. Again, the 3rd order intercept point is provided to the model as specified and the 2nd order intercept point and 1dB compression point cannot be parameterized. For the other parameters the same applies as for the LNA.

The *baseband filter* is realized as a Butterworth filter with lowpass characteristic (see Section 7.3.6). The model describes a 5th order filter with corner frequency of 9 MHz as specified.

The *baseband amplifier* uses the same model as the LNA. The gain is provided by a manually determined power budget calculation and presented as a fixed parameter value to the model.

Concerning the *frequency synthesizer* a further refinement in the model is necessary. The modeled PLL is a hierarchical composition of basic building blocks from the RF library. We use here a similar block diagram to that in

Section 7.5. Additional ideal splitter blocks are introduced in order to split the signal without any loss due to impedance mismatch. For the intended application of the PLL as a frequency synthesizer, the output *outclk* is relevant, whereas other applications such as FM demodulators require the output *out* after the loop filter.

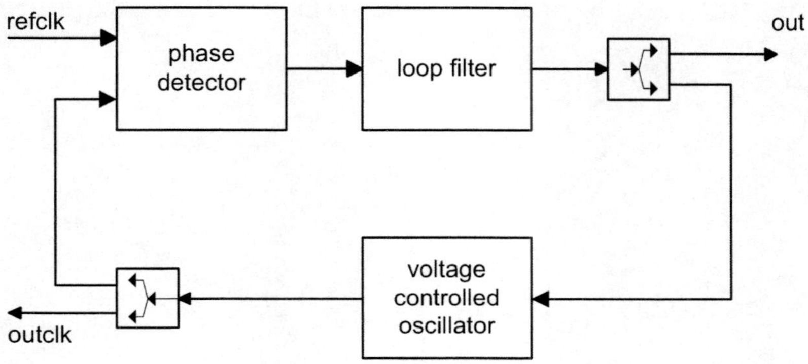

Figure 9-4. Block diagram of the modeled PLL

The center frequency of the PLL can be parameterized with the free running frequency of the VCO and is set to 2.6 GHz. Also, the initial phase offset and the signal power of the local oscillator can be adjusted in the VCO of the PLL.

Phase noise and orthogonality error are not included here. An oscillator model including phase noise is used in Chapter 10 for RF analyses of the WLAN receiver. The VHDL-AMS model of the PLL in Figure 9-4 can be found on the CD-ROM.

The PLL is fed by another VCO that has the same center frequency of 2.6 GHz. At the output of the PLL a *phase shifter* to splits the local oscillator clock into in-phase (I) and quadrature (Q) components. A simple ideal phase shifter model has been implemented that works at the specified LO frequency.

Other elements of the receiver include the *highpass filter* between the stages. Here we use simple capacitors that, together with the input impedance of the following stage, form a first-order highpass filter. A corner frequency of 1 GHz has been chosen to suppress DC offsets effectively.

The *splitter* block in the receiver is again ideal and identical to the one used for the PLL. Ideal in this sense means that the input signal is identically transferred to both outputs without any loss. To overcome the effect that the power is doubled in this case, which is unrealistic, further model refinements are necessary. The model is sufficient if only the in-phase path of the signal is examined.

In order to test the receiver a simple *transmitter* model needs to be established. Figure 9-5 shows the structure of the modeled transmitter. The dashed line part of the diagram is not modeled; it is drawn here only for completeness and to show the analogy to the receiver structure. The intended test simulation uses the in-phase channel (I) only. Therefore the phase shifter is also neglected.

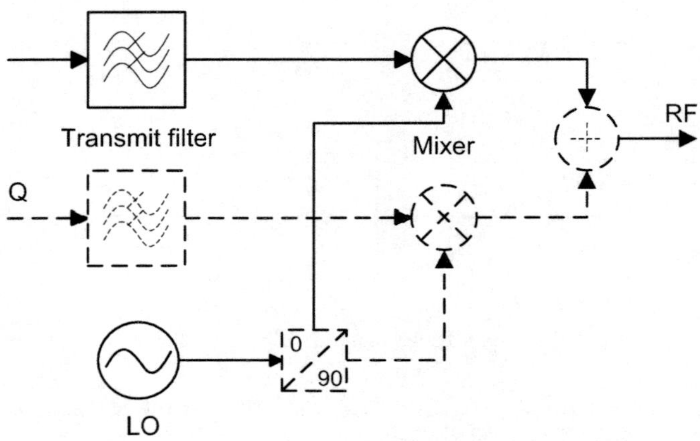

Figure 9-5. Block structure of the modeled transmitter

The upconversion of the transmitter signal to the passband of 5.2 GHz is done here in a single step without using an intermediate frequency.

A simple first-order lowpass filter with corner frequency of 8 MHz forms the transmit filter. It does not fulfill the Nyquist criterion, so intersymbol interference may distort the signal. But for first test runs of the receiver it might suffice. The VHDL-AMS model of the transmitter can be found on the CD-ROM.

As for the receiver only the RF section of the transmitter is modeled. Source and channel coding of the signal to be transmitted belong to baseband digital signal processing and are not included in this simulation. Specialized system simulators are available to develop this section. Nevertheless, since we use the mixed-signal language VHDL-AMS for modeling, the digital part of the circuit can be seamlessly integrated as VHDL blocks into the overall simulation.

To demonstrate this procedure, a pseudorandom binary source is used as the input signal. It is described completely in VHDL and can be used to represent any baseband digital signal processing block in VHDL.

9.4 Example Calibration

At this point of the design flow the initial specification of the receiver is subdivided into basic building blocks. The respective behavioral models in VHDL-AMS are basically developed from theory based on the system specification. Further refinement is necessary in order to improve the behavioral models in the way of accuracy and to fit them to circuit level models.

In this section an algebraic equation is fitted to a chosen characteristic that is extracted from a circuit level model. The equation can be used to describe the behavioral model. This step introduces an abstraction to reduce the simulation effort of the behavioral model compared to the circuit level model. Simulation effort is determined by the simulation time and the computational power that is required to execute the model. Furthermore the behavioral model should not strongly deviate from the extracted characteristic.

In the WLAN receiver model three amplifiers are specified, one low-noise amplifier at the input and two baseband amplifiers (I- and Q-channel) at the output of the receiver. The circuit level model *lnaSimple* is chosen from the Cadence library. This model is assumed to be the circuit level model of the baseband amplifiers. As an example characteristic the operating voltage dependency of the power gain is chosen. The nominal operating voltage of *lnaSimple* is 15 V. The input frequency is set to 1 MHz because the input signal of the baseband amplifier is a downconverted OFDM signal.

The circuit level model *lnaSimple* is simulated using SpectreRF. The power gain is plotted versus the operating voltage in a range from 0 V to 20 V. Figure 9-6 shows the simulation results.

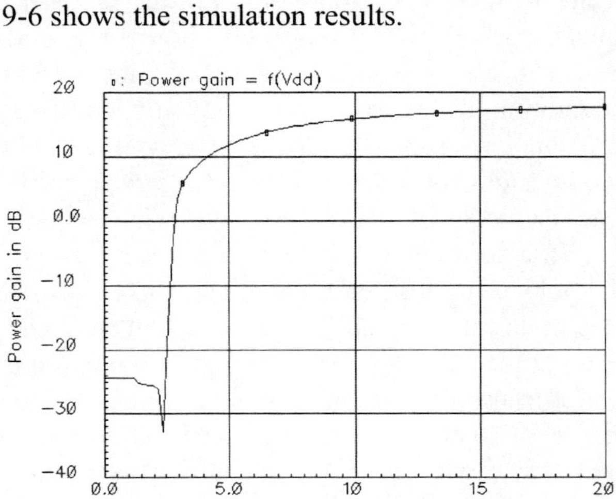

Figure 9-6. Power gain versus operating voltage of lnaSimple

The characteristic obtained can be handled by several curve fitting algorithms provided by tools like MATLAB or Mathematica. In this case MATLAB is used. The MATLAB plot in Figure 9-7 shows the characteristic obtained from the circuit level model.

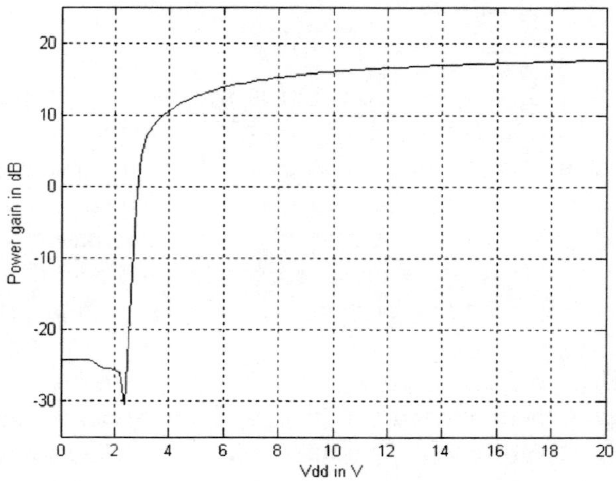

Figure 9-7. Power gain versus operating voltage in MATLAB

The range of the operating voltage that is used for fitting by a polynomial equation must be selected carefully. Close fitting of the complete curve results in a high degree of the polynomial and consequently leads to a high simulation effort of the model. Therefore only the region (12 V to 18 V) around the operating point of 15 V is chosen. In this region the dependency of the power gain from the operating voltage is linearly approximated. A strong deviation from the linear region can only be seen in the circuit characteristic below 4 V, which is far from the nominal operation point of 15 V. The resulting deviation from the circuit model must be considered with respect to the required accuracy of the relevant parameter in each individual case. The accuracy can be enhanced by increasing the degree of the polynomial or by using a different function for the algebraic equation.

MATLAB provides the function *polyfit(x,y,n)* where x and y are the data vectors of the plot (operating voltage and power gain) and n is the degree of the polynomial. Because the region considered is assumed to be linear, n is set to 1. This is also called linear regression. The output of the *polyfit* function is a row vector of the polynomial coefficients, in this case:

$$p = \begin{pmatrix} 0.1525 & 14.8412 \end{pmatrix}$$

Thus, a dependency of the power gain from the operating voltage of approximately 6.5 dB/V is observed. The algebraic equation to describe the chosen characteristic for the behavioral model is

$$y = 0.1525 \cdot x + 14.8412$$

where y is the power gain and x is the operating voltage. For the behavioral model the equation is valid in the range $12\text{V} < x < 18\text{V}$.

Outside of this region the behavioral model can still be used, but it will not reproduce the circuit behavior. A warning could be issued if the model range has been exceeded.

The statistic R-square is a measure for the accuracy of the model fit and represents the variation of the data. R-square can be determined easily by the Curve Fitting Toolbox that is provided by MATLAB. The range of R-square is between 0 and 1, with the latter value representing a perfect fit. In the example R-square is 0.9906 for the considered linear region which indicates a suitable fitting of the behavioral model to the circuit level model.

The MATLAB plot in Figure 9-8 shows the original curve and the fitted curve in a single window.

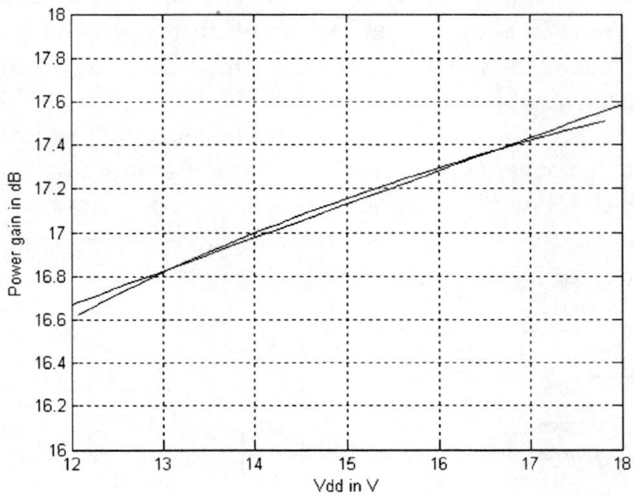

Figure 9-8. Linear regression of the operating point region

9.5 Example Verification

Verification of the WLAN receiver example is done using hierarchical composition of the modeled subblocks in terms of a bottom-up verification. Usually the subblock models have been individually calibrated and optimized using circuit models. After combining the optimized subblock models the overall system behavior is simulated. It is then possible to analyze whether the overall design goals and performance measures are met, bearing in mind the limited accuracy of the behavioral models.

For system level verification of the receiver design example we use the behavioral models that were described in Section 9.3. A system level test-bench for the receiver should include:

- Complete system level model for the design under test (DUT)
- Signal source to stimulate the DUT
- Signal processing block to adapt the source signal to the needs of the DUT
- Analysis blocks

Once a test-bench with these elements is established individual blocks can be replaced by more accurate ones or by circuit level implementations of the same block. Thereby the influence of this block on system level performance measures can be explored without needing to simulate the whole design at circuit level.

The treated receiver design example is completed with a binary source, a digital-to-analog converter (DAC) and the test transmitter to form the system level test-bench. Strictly speaking this is a pure RF test-bench since all baseband signal processing is neglected. Figure 9-9 shows the test-bench configuration to verify the receiver design example.

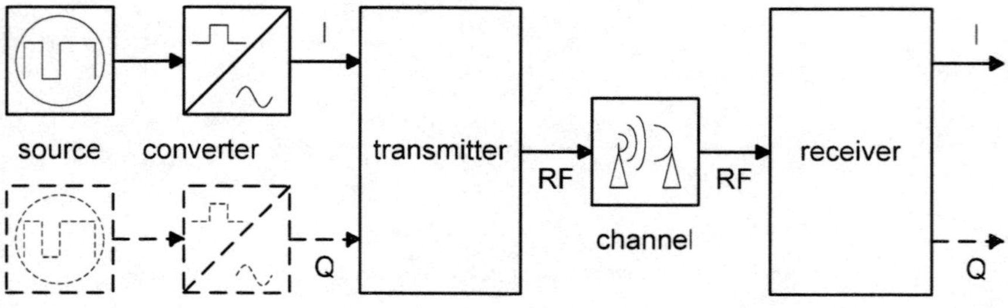

Figure 9-9. Testbench for system level verification

For the sake of simplicity only the in-phase channel (I) of the transmission system is considered in this example. A pseudorandom binary source (PRBS) supplies the input signal. The bit_vector-output of the PRBS is fed into a digital-to-analog converter (DAC) from the RF library yielding a time continuous representation of the binary sequence. This signal is to be transmitted over the channel. The simple transmitter model introduced modulates the signal on the carrier. The channel itself is, for the purpose of this example, not modeled, that is, it is an ideal channel without any loss. The receiver is used as described earlier.

The VHDL-AMS model of the complete test-bench can be found on the CD-ROM.

Only a few bits are simulated since the simulation takes a very long time. This is mainly because of the very high carrier frequency of 5.2 GHz. This frequency has to be simulated because we use real passband models instead of complex baseband representations. For overall system parameters like bit error rate estimation, complex baseband models are better-suited (see Section 4.3) whereas our passband simulation allows us to study harmonic distortion due to intermodulation at the nonlinearities.

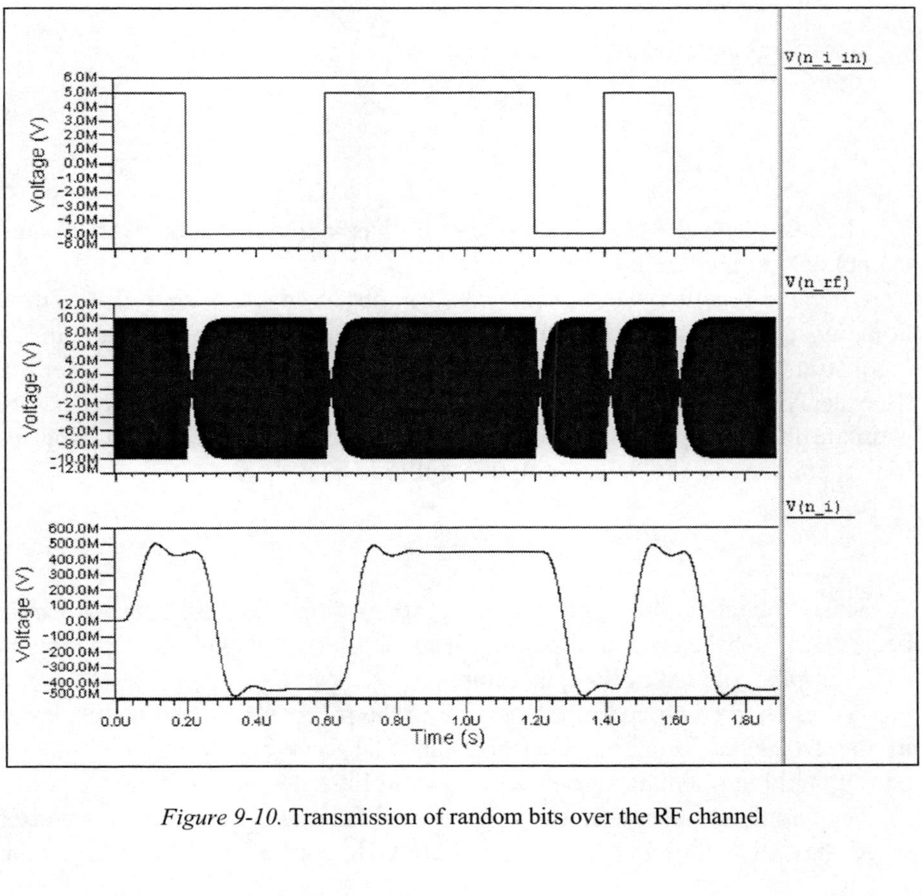

Figure 9-10. Transmission of random bits over the RF channel

Figure 9-10 shows the simulation of the receiver design example in the testbench. The signal waveforms are shown in the time domain. 20 Bits were simulated at a rate of 5 kBit/s. The simulation took almost 30 minutes to complete over 2 µs real-time on a Sun Ultra 250 with a 400 MHz clock. A complete circuit level passband simulation of the same system would take days to simulate.

The binary sequence is generated by a linear feedback shift register (LFSR) of length $n=3$, that is, it has a period of $2^3-1=7$. At the input of the transmitter (node *n_i_in*) the signal is already converted into the analog domain. It has a power level of

$$p_{dBm}(n\_i\_in) = -30dBm$$

which corresponds to

$$v_{pp}(n\_i\_in) = 10mV$$

at R=50 Ω according to the formula

$$v_{pp}(n\_i\_in) = \sqrt{10^{\frac{p_{dBm}(n\_i\_in)}{10}} \cdot 2R}$$

The signal on the channel (node *n_rf*) has a frequency of 5.2 GHz and cannot be resolved in this diagram.

After the receiver the waveform *n_i* is observed. As shown in the time domain representation (Figure 9-10) the original binary pattern can be reconstructed after transmission. The whole transmission system inserts a time delay of approximately 0.1 µs. A subsequent baseband section has to estimate this delay in order to sample the signal correctly in an analog-to-digital converter (ADC). The output amplitude of

$$v_{pp}(n\_i) = 0.889V$$

nearly matches the specified value of 1 Vpp. The difference occurs because of a phase error in the receiver, which leads to spurious components of the in-phase signal in the quadrature path.

The method of processing the signal in the receiver can also be displayed in the frequency domain. Therefore an FFT was performed with input, output, and intermediate signals, as shown in Figure 9-11.

The input frequency (node *n_rf*) lies at 5.2 GHz. Both downconverter stages have the same LO frequency of 2.6 GHz applied, which is half of the carrier frequency. Thus after the first mixer stage the signal *n_if* can be found at both 2.6 GHz and 7.8 GHz. The second mixer stage converts the signal down to baseband (*n_bbi*). Other (partly mirrored) images of the required signal lie at 5.2 GHz and 6 GHz. These images are rejected by the bandpass filter at the receiver output. Therefore, the output signal *n_i* only contains the required baseband component. It can be observed that the gain of the amplifiers and the mixers raise both the signal and the noise level.

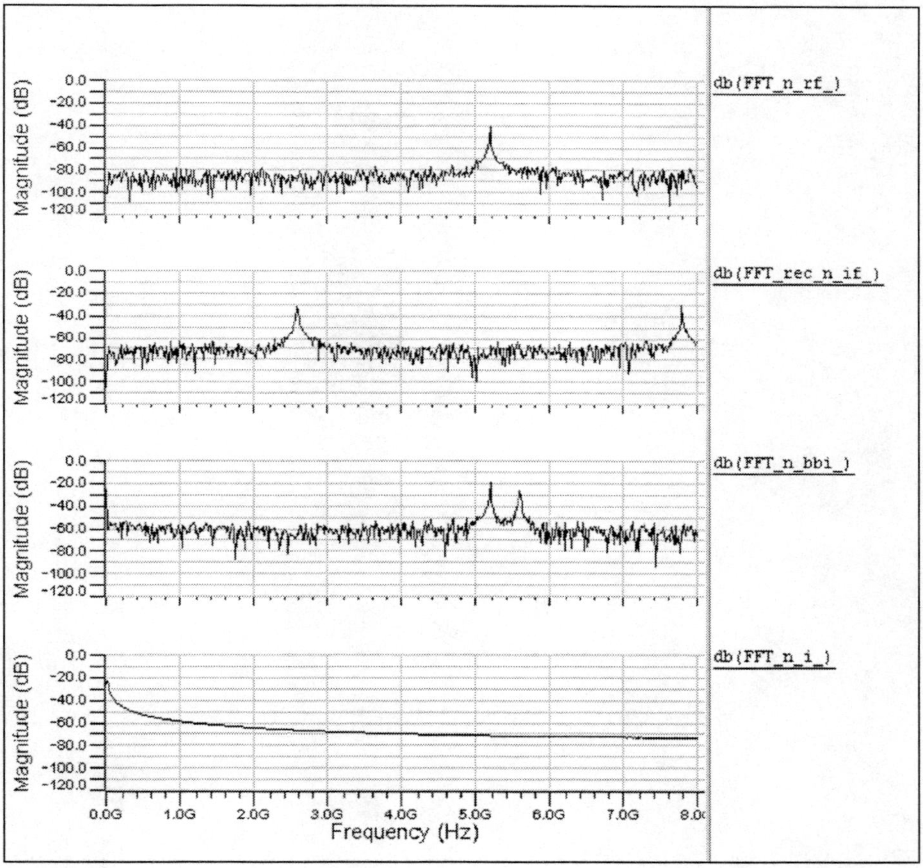

Figure 9-11. Signal transformation in the frequency domain

Chapter 10

MODELING OF ANALOG BLOCKS IN VERILOG-A

10.1 Introduction

This chapter deals with the modeling of analog blocks and systems in *Verilog-A* and their simulation in the *Cadence Analog Design Environment* (ADE).

Verilog-A is a high-level hardware description language standard which is a subset of the mixed-signal modeling language Verilog-AMS. It is used to describe the structure and behavior of analog systems. The analog statements of Verilog-A can be used to describe a wide range of systems, such as electrical, mechanical, fluid dynamic, and thermodynamic systems. To specify the behavior of individual modules, mathematical relationships between their input and output signals can be defined.

The simulator *Spectre* is the analog circuit simulation tool from Cadence ADE. The RF option, *SpectreRF* [Cad03a], provides specific simulation algorithms for the analysis and characterization of RF components, which may include frequency conversion effects. The components can be described at circuit level using netlists or schematics. Behavioral models can be used to describe higher levels of abstraction of the design. Therefore the simulation can be accelerated to quickly evaluate different system architectures. Spectre supports two behavioral modeling languages:

- *SpectreHDL,* which is a Cadence specific non-standard language
- Verilog-A, which has been standardized by *OVI (Open Verilog International)*

Because of the standardization and the interoperability, Verilog-A is widely used. Section 10.2 is dedicated to the development of behavioral models in Verilog-A. Section 10.3 provides an overview of the behavioral models of the Cadence library *rfLib*. Their usage is illustrated on the simulation of a WLAN receiver in Section 10.4.

10.2 Writing Custom Behavioral Models

This section describes the development of user-defined behavioral models. In the first part a short overview of Verilog-A is given with respect to the requirements of RF modeling. An example shows the implementation and simulation of an RF model using Verilog-A, Cadence ADE, and SpectreRF. In the last section a tutorial is introduced where the reader can learn how to create a behavioral model in Verilog-A.

10.2.1 Verilog-A principles

Verilog-A (see [Acc04] and [Cad03c]) uses two different approaches to describe analog behavior. There are *conservative systems* (Kirchhoff's laws) and *nonconservative systems* (signal flow). Both systems can be described in terms of mathematical equations or by interconnected systems. These concepts can be realized by means of ports, nodes and branches. Furthermore the description of time continuing systems using *differential algebraic equations* (DAE) is possible.

Conservative systems always represent two quantities, namely potential and flow natures. In the electrical domain this leads to the representation of voltages and currents for example. Signal flow systems only represent potential natures. Natures define the characteristics of physical dimensions, tolerance requirements and access functions. The disciplines in Verilog-A correspond to the natures in VHDL-AMS. Verilog-A contains operators for the modeling of noise and the modeling of frequency and transfer functions.

Table 10-1 provides a short outline of functions which are available in Verilog-A. Among other application areas they are useful to model the behavior of high-frequency systems.

Table 10-1. Verilog-A functions useful for RF modeling

Modeling task	Verilog-A functions
Frequency and transfer functions	laplace_zp, laplace_zd, laplace_np, laplace_nd, zi_zp, zi_zd, zi_np, zi_nd, delay, idt, ddt
Large-signal noise	$dist_normal, $rdist_normal, … (a total of 7 different distributions)
Small-signal noise	white_noise, flicker_noise, noise_table
Event detection	cross, timer

The basis for behavioral and structural descriptions are *modules*. The module *lp_filter* (see listing below) shows the basic structure of a Verilog-A module. Port and signal declarations are used to describe the interface of the module. In the *parameter* declaration *electrical* or other physical values are defined. Other declaration types like *real* or *integer* are also possible. The underlying description of the module behavior is done with a structural or a behavioral part. All description types, system types, and disciplines can be combined in a single Verilog-A model.

```
module lp_filter (sig1, sig2, gnd);
   inout sig1, sig2, gnd;
   electrical sig1, sig2, gnd;
   parameter real R = 1k;
   parameter real C = 1u;
   // structural description, see lp_filter_str
   // behavioral description, see lp_filter_beh
endmodule
```

In a structural description several modules can be instantiated and connected. In this case the design becomes hierarchical and facilitates the top-down design process. The module *lp_filter_str* shows the structural description of a lowpass filter, where the modules of the resistor and the capacitor are instantiated.

```
module lp_filter_str (sig1, sig2, gnd);
   inout sig1, sig2, gnd;
   electrical sig1, sig2, gnd;
   res res_inst (.r_in(sig1), .r_out(sig2));
   cap cap_inst (.c_in(sig2), .c_out(gnd));
endmodule
```

A behavioral description contains the mathematical relationships between input signals, output signals and parameters. For that purpose Verilog-A contains a rich set of analog operators and functions. The module *lp_filter_beh* shows the behavioral description of the lowpass filter using the *ddt* time derivative operator.

```
module lp_filter_beh (sig1, sig2, gnd);
   inout sig1, sig2, gnd;
   electrical sig1, sig2, gnd;
   parameter real R = 1k;
   parameter real C = 1u
   analog begin
      I(sig1, sig2) <+ V(sig1, sig2)/R;
      I(sig2, gnd) <+ ddt(V(sig2, gnd)*C);
   end
endmodule
```

This section does not provide a complete overview of Verilog-A. Examples corresponding to the different assignment and control statements, data types or lexical conventions are omitted here. The following sections describe more comprehensive examples and describe Verilog-A in a more detailed manner.

10.2.2 LNA modeling example

In this section analog behavioral modeling with *Verilog-A* and the *Cadence ADE* is shown at an example of a *low-noise amplifier* (LNA). The functional description and important main characteristics are already described in Section 7.3.1 as a VHDL-AMS model.

Verilog-A model of a LNA

The VHDL-AMS model is transferred to Verilog-A and additionally includes the modeling of noise. A few structural and mathematical deviations were also made. Table 10-2 gives a short overview of the model parameters. The model parameter *IP3* can be directly calculated with special RF analyses and postprocessing capabilities provided by *SpectreRF*.

Table 10-2. LNA parameters

Parameters	Unit	Value	Description
gain	dB	6.8	Power gain
ip3	dBm	1.9	3rd order intercept point
fnoise	dB	5	Noise figure
fg	Hz	10M	3dB frequency
rin	Ω	50	Input resistance
rout	Ω	50	Output resistance

Model implementation

Since this model implementation of the LNA is derived from the VHDL-AMS model, a repeated description is omitted. A more detailed description of a Verilog-A module can be found in the Section 10.2.3 where a mixer is implemented.

```
`include "constants.h"
`include "discipline.h"

module amp (in, out, gnd);

  inout in, out, gnd;
  electrical in, out, gnd;
  electrical p1, p2, p3;
```

```
  parameter real gain = 6.8;          // Gain in dB
  parameter real rin = 50;            // Input resistance
  parameter real rout = 50;           // Output resistance
  parameter real fg = 10M;            // 3 dB frequency
  parameter real fnoise = 5;          // Noise figure in dB
  parameter real ip3 = 1.9;           // IP3 in dBm

  real gain_lin, ip3_lin, noise, a, b, inmax, outmax;

  analog begin
    @ (initial_step) begin
      gain_lin = pow (10, gain/10);
      ip3_lin = sqrt ((pow (10, ip3/10)) * 2*rin*0.001);
      noise = 4*`P_K*(pow(10,fnoise/10)-1)*$temperature*rin;
      a = sqrt((gain_lin*rout)/rin);
      b = (4*a)/(3*ip3_lin*ip3_lin);
      inmax = sqrt(a/(3*b));
      outmax = (2*a*inmax)/3;
    end

    // noise source
    V(in,p1) <+ white_noise (noise, "noise");

    // input resistance
    V(p1,gnd) <+ I(p1,gnd) * rin;

    // frequency response
    V(p2,gnd) <+ laplace_nd(V(p1,gnd), {1}, {1, 1/`M_TWO_PI/fg});

    // nonlinear characteristic
    if (abs(V(p2,gnd)) < inmax)
       V(p3,gnd) <+ 2 * (a - b*V(p2,gnd)*V(p2,gnd)) * V(p2,gnd);
    else if (V(p2,gnd) > 0)
       V(p3,gnd) <+ 2 * outmax;
    else
       V(p3,gnd) <+ -2 * outmax;

    // output resistance
    V(out,p3) <+ I(out,p3) * rout;

  end

endmodule
```

Simulation results

This part describes the simulation results of the LNA model. Most of the analyses (for example *DC analysis*, *AC analysis*) are conventional analyses used in Spectre. In contrast the *IP3* and the *1dB CP* measurements are executed by RF analyses providing the required postprocessing of simulation data.

Figure 10-1 shows the *transient* simulation results. The input and the output signals are plotted in the time domain. The input signal is amplified by the model parameter gain.

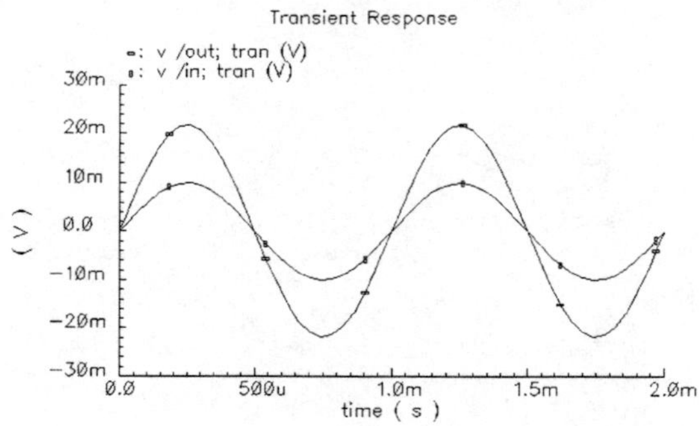

Figure 10-1. Transient response of the LNA

Figure 10-2 depicts the nonlinearities modeled in the corresponding section of the Verilog-A model. The gain of the LNA is limited by the parameter *outmax* which represents an internal variable.

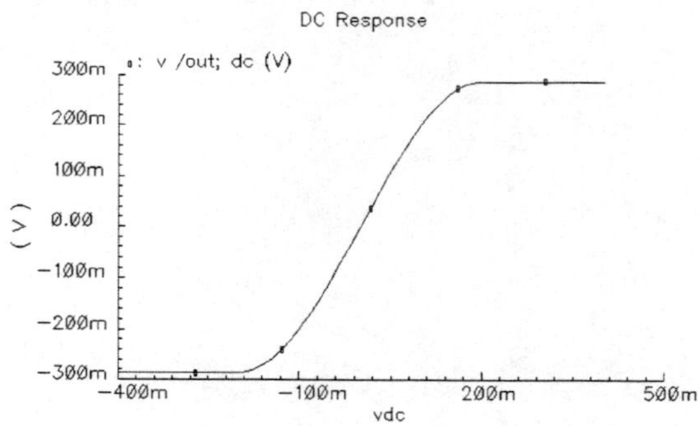

Figure 10-2. DC response of the LNA

The frequency response with the corner frequency *fg* = 10 MHz is modeled using the *laplace_nd* operator and is performed by an *AC analysis*. In Figure 10-3 the cross marker *A* labels the *fg* and 3 dB loss. The gain of 6.8 dB can be read off the graph. Instead of the conventional AC analysis the *Periodic AC* (PAC) analysis may be used for the LNA. The AC analysis would be unsuitable for the mixer due to the frequency conversion effects.

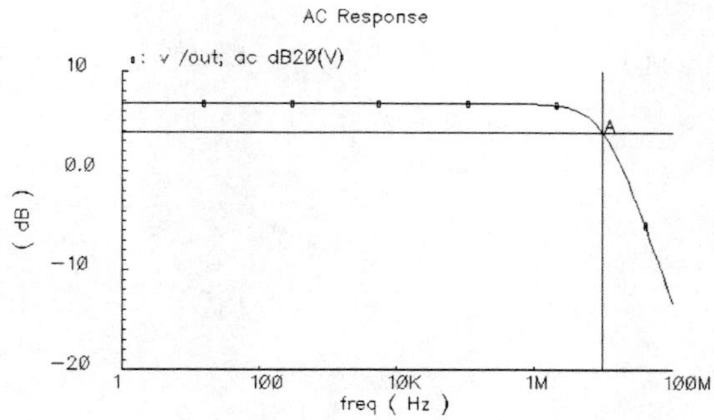

Figure 10-3. AC response of the LNA

Figure 10-4 shows the *noise figure* (NF) of the model according to the parameter *fnoise* = 5 dB. The noise figure is the ratio of SNR_{in} to SNR_{out} and is usually expressed in terms of dB. Using SpectreRF the noise figure could be calculated with the *PNoise* analysis, which is a periodic small-signal analysis. In circuit level models the noise figure depends on the frequency.

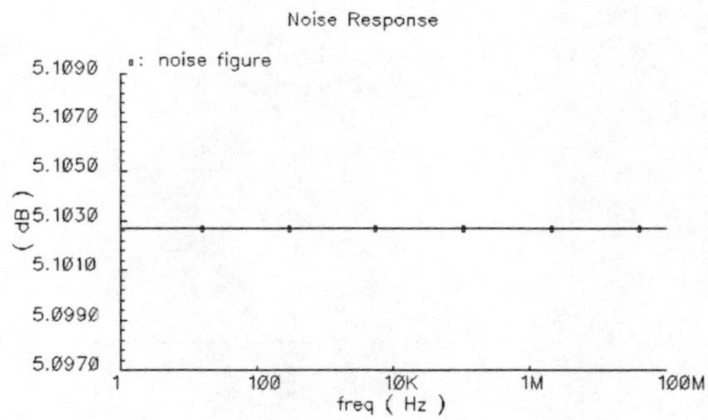

Figure 10-4. Noise figure of the LNA

The next LNA parameter is the *3rd order Intercept Point* (IP3). The measurement results are achieved by a combined *PSS/PAC* analysis using two tones of 1 kHz and 1.1 kHz. Figure 10-5 depicts the IP3 of approximately 1.9 dBm. A detailed explanation of the IP3 can be found in Section 11.2.

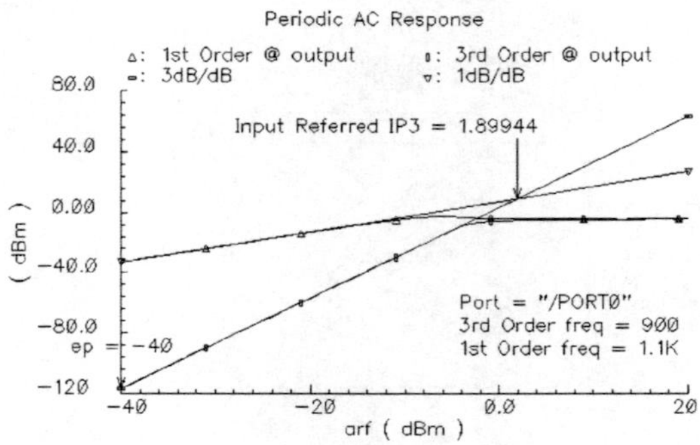

Figure 10-5. IP3 of the LNA

Figure 10-6 depicts the *1dB Compression Point* (1dB CP) which has a distance of approximately 10 dBm to the IP3. The 1dB CP is the point where the output power falls 1 dB below the 1dB/dB curve. Beyond the 1dB CP the model is in the saturation region.

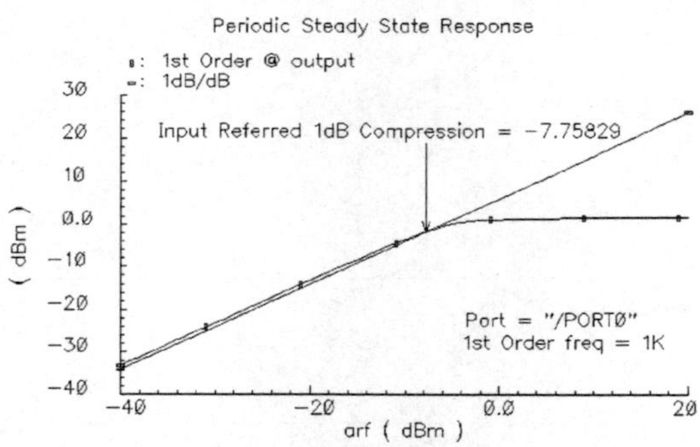

Figure 10-6. 1dB CP of the LNA

10.2.3 Creating a Verilog-A model

Objective

A Verilog-A model of a *mixer* shall be created. The inputs are *p_rf* (RF signal), *p_lo* (local oscillator signal) and the output is *p_if* (intermediate frequency signal). Important parameters of the mixer are:

- Port impedances each of 50 Ω
- Gain of 0 dB
- Corner frequency of 1 GHz
- Noise figure of 5 dB
- IP3 of -30 dBm

The following formulas are used to describe the behavior of the model. They are partially described in Section 7.3.1.

- Conversion of logarithmic power gain into a linear power gain:

$$gain\_lin = 10^{\frac{gain}{10}}$$

- Conversion of the logarithmic *3rd order intercept point* (IP3) into a linear IP3 (refers to a single-tone signal), *r_rf* specifies the input impedance:

$$ip3\_lin = \sqrt{10^{\frac{ip3-30}{10}} \cdot 2 \cdot r\_rf}$$

- The small-signal noise is calculated by the following relationship using the *Boltzmann* constant *k*, the temperature *T*, and the variable *fnoise* which specifies the value of the noise figure:

$$noise = 4 \cdot k \cdot (10^{\frac{fnoise}{10}} - 1) \cdot T \cdot r\_rf$$

- The nonlinearity is modeled using the depicted characteristic which includes 3rd order effects:

$$v_{out} = a \cdot v_{in} - b \cdot v_{in}^{3}$$

- The values *a* and *b* represent the coefficients of the nonlinear characteristic, *r_if* specifies the output impedance:

$$a = \sqrt{gain\_lin \cdot \frac{r\_if}{r\_rf}} \quad b = \frac{4 \cdot a}{3 \cdot ip3\_lin^2}$$

- Frequency response which is implemented in numerator-denominator form:

$$H(s) = \frac{1}{1 + \frac{1}{\omega_g} s}$$

Proposal for solution

The head of the mixer module includes the interface declarations (for example inout ports of type electrical). For the internal usage electrical signals are additionally declared. All parameters for the mathematical description and variables are defined next.

```
`include "constants.h"
`include "discipline.h"

module mixer (rf_in, lo_in, if_out, gnd);

  inout rf_in, lo_in, if_out, gnd;
  electrical rf_in, lo_in, if_out, gnd;
  electrical p1, p2, p3, p4;

  parameter real gain = 0;        // Gain in dB
  parameter real r_rf = 50;       // Input resistance RF Input
  parameter real r_lo = 50;       // Input resistance LO Input
  parameter real r_if = 50;       // Output resistance IF Output
  parameter real fg = 1G;         // 3dB frequency
  parameter real fnoise = 5;      // Noise figure in dB
  parameter real ip3 = -30;       // IP3 in dBm

  real gain_lin, ip3_lin, noise, a, b, inmax, outmax;
```

The parameters can also be changed in the properties form of the modules symbol view, which can be generated after Verilog-A creation in Cadence ADE.

The keyword *analog* introduces the analog section of Verilog-A modules. It defines the behavior as a procedural sequence of different statement types (see also [Acc04]) and is executed at every simulation point. The keyword *initial_step* generates a global event at the first simulation point in an analysis. In the listing necessary transformations are made to precalculate the specified parameters of the mixer into an internal representation. Logarithmic values are transformed into linear values for example. Here, the given formulas are used.

```
analog begin
  @ (initial_step) begin
    gain_lin = pow (10, gain/10);
    ip3_lin = sqrt (pow (10, (ip3/10)) * 2*r_rf*0.001);
    noise = 4*`P_K*(pow(10,fnoise/10)-1)*$temperature*r_rf;
    a = sqrt((gain_lin*r_if)/r_rf);
    b = (4*a)/(3*ip3_lin*ip3_lin);
    inmax = sqrt(a/(3*b));
    outmax = (2*a*inmax)/3;
  end
```

After the *initial_step* the behavior of the model can be described. These are for example noise, frequency response, and nonlinear characteristics.

The characteristics of the mixer are encapsulated into subblocks and connected within the Verilog-A module. Figure 10-7 gives an overview of the model.

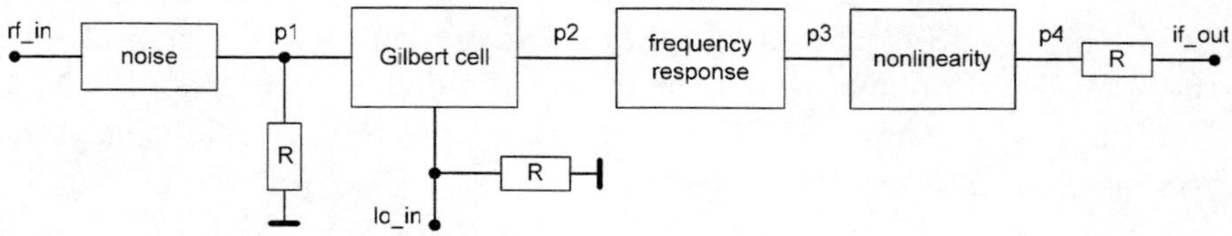

Figure 10-7. Overview of the mixer model

The input and output resistances are modeled using current-voltage relationships. The noise source is modeled using the *white_noise* function which is provided by Verilog-A. This function is only active during small-signal analyses, otherwise it returns zero. The frequency response is described with the *laplace_nd* function. Nonlinearity is modeled using the introduced characteristic and the limitation of input and output amplitudes. The simple model of a *Gilbert cell* multiplies the signals *p1* and *lo_in*. Compared to the LNA described in Section 10.2.2 only the second input *lo_in* and the model of the Gilbert cell are added.

```
    // noise source
    V(rf_in,p1) <+ white_noise (noise, "noise");

    // input resistance
    V(p1,gnd) <+ I(p1,gnd) * r_rf;
    V(lo_in,gnd) <+ I(lo_in,gnd) * r_lo;

    // Gilbert cell
    V(p2,gnd) <+ V(p1,gnd) * V(lo_in,gnd);

    // frequency response
    V(p3,gnd) <+ laplace_nd(V(p2,gnd), {1}, {1, 1/`M_TWO_PI/fg});

    // nonlinear characteristic
    if (abs(V(p3,gnd)) < inmax)
       V(p4,gnd) <+ 2 * (a - b*V(p3,gnd)*V(p3,gnd)) * V(p3,gnd);
    else if (V(p2,gnd) > 0)
       V(p4,gnd) <+ 2 * outmax;
    else
       V(p4,gnd) <+ -2 * outmax;

    // output resistance
    V(if_out,p4) <+ I(if_out,p4) * r_if;

  end

endmodule
```

Simulation results

With the following simulation examples the functionality of selected model characteristics are verified.

Figure 10-8 shows the frequency translation of the input signal (900 MHz, -50 dBm) according to the local oscillator (LO) signal (1 GHz, -10 dBm). The IF output shows the expected output frequencies (100 MHz, -76 dBm and 1.9 GHz, -83 dBm). For simulation the periodic steady state (PSS) analysis is used.

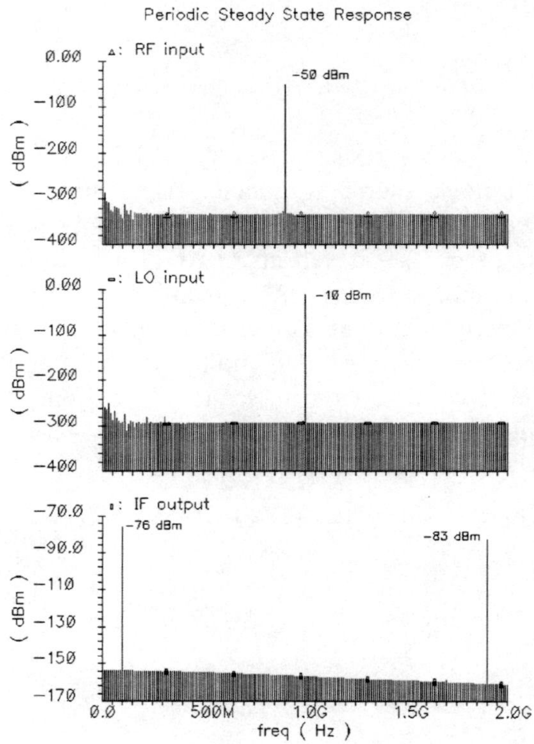

Figure 10-8. Frequency translation

Figure 10-9 depicts the *conversion gain* (-6 dB) and the 3dB corner frequency (1 GHz) of the mixer. A strong LO signal of 10 dBm, which is equivalent to 1 V at 50 Ω, is used in this combined *PSS/PXF* (PXF, Periodic Transfer Function) analysis. Since the conversion gain of this mixer also depends on LO power, the chosen value of 1 V neglects this effect. Furthermore the conversion gain originates from the signal split into downconversion and upconversion.

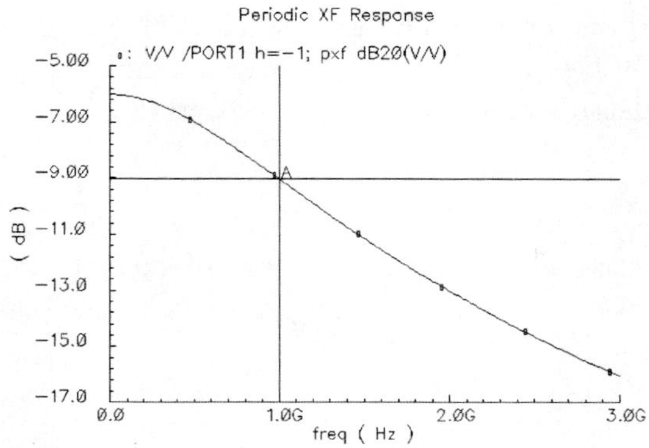

Figure 10-9. Conversion gain and 3 dB frequency

10.3 Overview of the Cadence Model Library rfLib

The modeling of systems and components can be simplified by using existing model libraries. To meet the aspects of reuse Cadence provides a number of libraries containing analog behavioral models, which can be applied especially in a *top-down* design flow. The library *rfLib* provides models dedicated to RF system design [Cad03a].

An overview of the *rfLib* is given in this section, subdivided into three parts. The first part of this section contains models of the most common RF building blocks used for *top-down design*. They can be building blocks for complex RF systems or executable specifications at the behavioral level. The models are described in Verilog-A and can be inserted into regular RF circuits for simulation. The required block parameters are translated into internal coefficients and equations that describe the relations between the voltages and currents at the connecting nodes.

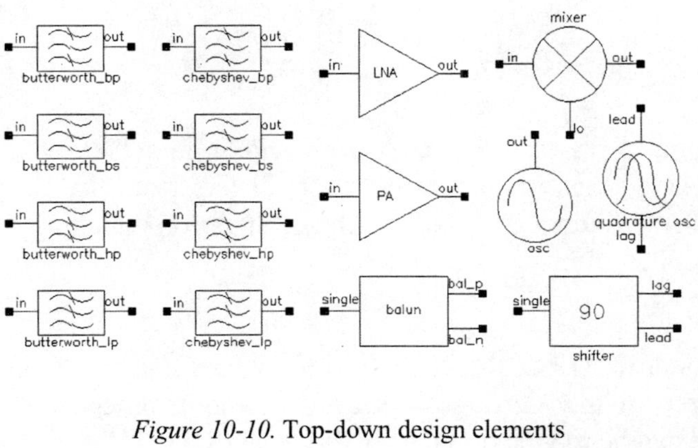

Figure 10-10. Top-down design elements

Figure 10-10 shows several top-down design elements: filters, balun, low-noise amplifier (LNA), mixer, power amplifier (PA), oscillator, quadrature oscillator, and phase shifter.

Filter

Filter properties are specified in the frequency domain. Lowpass, highpass, bandpass and bandstop filters are implemented, and each can be a *Butterworth* or *Chebyshev* type filter.

Balun

A balun is used in circuits that require single to differential signal transformation. Although in reality a passive network is used to realize the balun, this implementation employs a three-port network.

LNA

Low-noise amplifiers are commonly used in the receiver design to amplify the signal with a low noise figure. A typical low-noise amplifier has three sets of parameters: linear model, nonlinear model, and noise model parameters.

Mixer

Mixers are used for frequency translation in RF circuits. Basically a mixer has the following three sets of parameters: time-varying linear model, nonlinear model, and noise model parameters. The *rfLib* model describes the typical behavior of integrated mixers. The LO signal switches the input signal on and off. When the LO power exceeds the specified limit it is effectively clipped off.

PA

Power amplifiers are used in *RF transmitters* to achieve the high power output levels. The power amplifier model differs from the LNA in having greater power delivery capabilities with less stress on matching capabilities.

Oscillator

This model describes the essential information for an oscillator or local power source in the Verilog-A language. Among other important issues, phase noise can be modeled in the small-signal domain.

Quadrature oscillator

This oscillator is used in quadrature receiver design. A phase shifter is ordinarily used to generate the quadrature signal from one signal source. However, it is difficult to implement a phase shifter into a wide band model. A quadrature signal consists of two signals with a 90° phase difference but with identical noise and amplitude.

Phase shifter

In digital *RF system designs* the quadrature signal processing involves the phase splitting of high-frequency signals. The most common use of such components is to generate two signals that have a 90° phase difference based on one signal source.

Most of the itemized models are provided in two different versions for *passband* and *baseband signal* handling. In this second part of the section the baseband principle and the *equivalent baseband* models are briefly introduced.

The complex baseband simulation is a commonly used technique in system level simulation. The principle of complex baseband simulation is presented in Section 4.3. Starting with the release IC 4.4.6 Cadence also provides analog behavioral models for complex baseband signals.

As shown in Figure 10-11 each input and output signal of a baseband model has two components (pins) in contrast to a passband model. They represent a complex valued signal. It can be interpreted as amplitude and phase of a modulated carrier transformed from polar to rectangular coordinates. The *inphase* component is represented by the signal I and the *quadrature* component is represented by the signal Q. The transformation of a passband signal to its equivalent baseband representation is also described in Section 4.3.

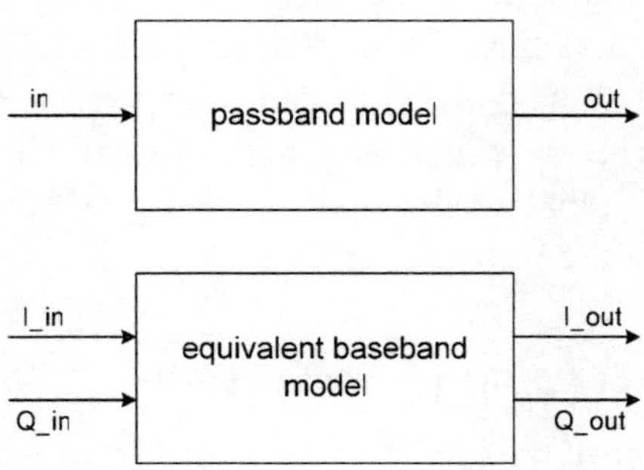

Figure 10-11. Equivalent baseband model

Complex baseband models can be used during the specification process of the *RF subsystem* to reduce the simulation time. Since these models deal with complex valued input and output signals, they cannot be connected directly to circuit level block models. If baseband models are connected to circuit level or behavioral passband models appropriate signal converters (for example an IQ-modulator) must be inserted. The signal converters combine the baseband signal with the specified carrier frequency or vice versa. Thereby the signal representation changes.

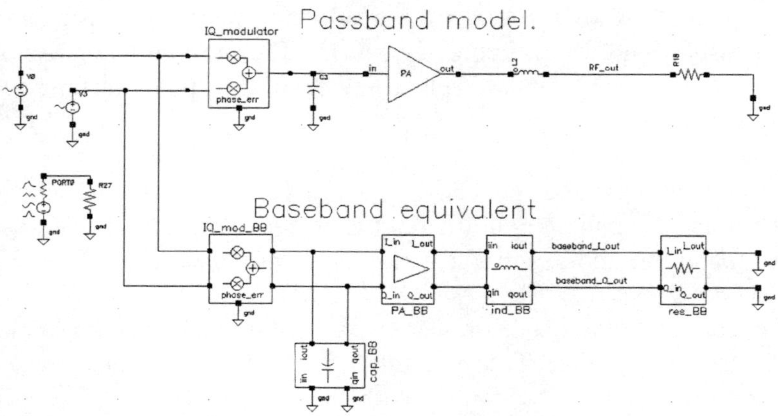

Figure 10-12. BB_testbench circuit

Figure 10-12 shows two equivalent circuits, the first modeled using passband models and the second using equivalent baseband models. The same baseband signal drives both circuits. In the passband circuit the baseband signals are first mixed up with the passband using the model *IQ_modulator*. In the equivalent baseband model *IQ_mod_BB* only modulation and mixing effects like nonlinearities are added to the signal.

This last part of this section describes elements used in test-benches representing sources and sinks for system simulation. Three different sources of digital modulated signals are provided to test RF circuits with practically used input signals. The *eye-diagram* generator provides a special signal plot, which allows us to visualize the quality of digital modulated signals. The library *rfLib* contains the following elements:

CDMA Signal Source (CDMA_reverse_xmit)

The *CDMA* (Code Division Multiple Access) signal source generates a reverse-link (handset-to-base-station) signal conforming to the IS-95 standard. The modulation is an offset *QPSK* (Quadrature Phase Shift

Keying) with a symbol rate of 1.2288 Mega-symbols/s and a sample rate of 4.9152 Mega-samples/s. Two separate 16-bit pseudo-noise generators generate the I and Q spreading sequences operating at the sample rate. Each sequence is filtered with a 48-tap FIR filter.

GSM Signal Source (GSM-xmtr)

Using *GMSK* (Gaussian Minimum Shift Keying) modulation the *GSM* (Group Special Mobile) source generates a signal conforming to the *GSM mobile communication* standard. The first part of the GSM signal source is a random binary generator with a bit rate of 270833.333 bits/s. The following FIR filter is a Gaussian filter implemented with 32 taps and signal gain. In the next step the signal is integrated using a modulo 2π integrator and split into I and Q channels.

π/4-DQPSK Signal Source (pi_over4_dqpsk)

The *π/4 differential QPSK* baseband signal source generates random binary data with a bit rate of 48.6 kbits/s. The data is converted from serial to parallel (2 bits). A phase state coder maps pairs of bits to phase shift and the following differential encoder shifts the symbol phase. Next the I and Q channels are filtered. The FIR filter used is implemented with 64 taps.

The constellation diagram in Figure 10-13 depicts how the differential encoded binary signal is mapped to the carrier phase.

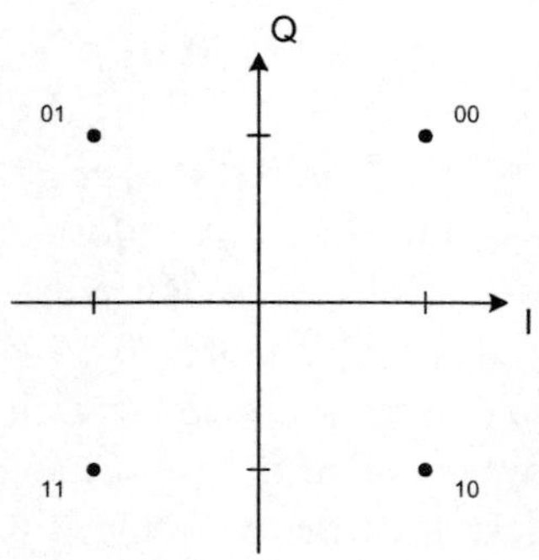

Figure 10-13. Constellation diagram

Eye-diagram generator

The *eye-diagram* is a widely used plot format to evaluate the quality of digital modulated signals. The signals are plotted over a period of an integer multiple of the symbol duration. This functionality is realized in the waveform viewers of system level simulators as postprocessing. To provide such functionality in the *Cadence Analog Artist* waveform viewer an appropriate generator is delivered. The input to the eye-diagram generator is the I or Q component of a complex baseband signal. The eye-diagram generator provides two outputs labeled "y-axis" and "x-axis". The eye-diagram is generated by plotting the y-axis output against the x-axis output in the Analog Artist waveform viewer. An example of an eye-diagram is shown in Figure 10-14. To generate eye-diagrams the transient simulation must be used.

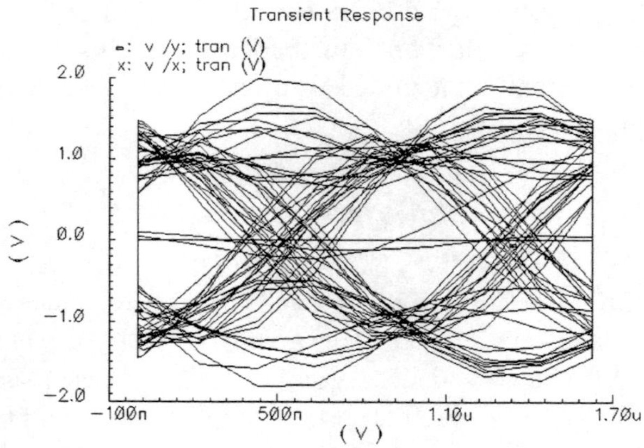

Figure 10-14. Eye-diagram

10.4 Modeling and Simulation of a WLAN Receiver

This design example is based on a complex system level architecture for a wireless high-speed data transmission conforming to the *WLAN standard IEEE 802.11a*. It has been simplified to concentrate on the main parameters and to keep the example practical. Figure 10-15 shows the WLAN receiver corresponding to a double conversion receiver architecture.

This example is the same as that used in Chapter 9. The input signal and block specification are described there. While it is treated there extensively at system level, we focus here on circuit level simulations.

The task of the WLAN receiver is to receive a broadband signal at radio frequency from the antenna and to transform the signal (for example in frequency, power and phase) for digital baseband processing. The incoming signal is assumed to be an *OFDM* (Orthogonal Frequency Division Multiplex) signal. It is first amplified by an LNA and then downconverted to baseband at two mixer stages working at the same local oscillator (LO) frequency.

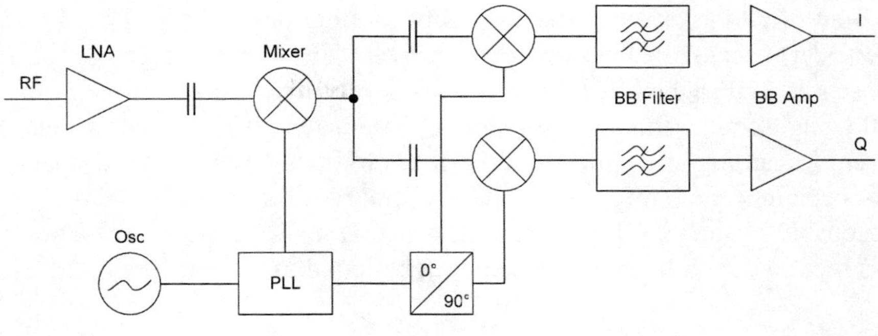

Figure 10-15. Block diagram of the WLAN receiver

Between the mixer stages the signal is split into I and Q signals using a 90° phase shifter for oscillator frequency. After that the required channel is filtered by a bandpass filter and amplified by a baseband amplifier.

10.4.1 WLAN receiver modeling using Cadence libraries

Figure 10-16 shows the realized WLAN receiver model, which basically consists of RF components located in the *rfLib* library [Cad03a]. The models used like LNA, mixer, filter or oscillator are already introduced in Section 10.3. Further elements of the receiver are sources and sinks using *psin* and *gnd* elements provided by the *analogLib* library.

In order to simplify the receiver model and its simulation, the specified *PLL* is not realized. Instead an oscillator producing a fixed frequency of 2.6 GHz and *phase noise* is used to drive the first mixer and the phase shifter. The phase shifter splits the oscillator signal and feeds both mixers of the second stage with a phase difference of 90°.

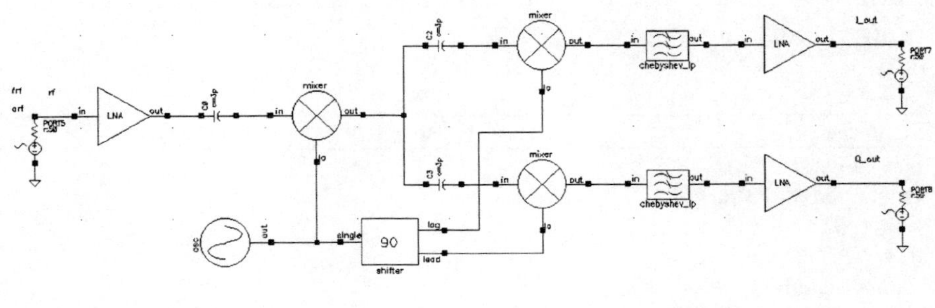

Figure 10-16. WLAN receiver modeled using Cadence ADE

The specified bandpass filter and baseband amplifier (AGC) are replaced by a lowpass filter and an LNA with fixed gain. The capacitors (*analogLib*) in front of all the mixer blocks are useful to filter out DC offsets and *flicker noise* (1/f) because of their highpass behavior. The value of each capacitor is chosen as 3 pF, which results in a 3 dB corner frequency of 1 GHz. In the following section the most important parameters of all components are given. Regarding all applied models, it has to be mentioned that unspecified block parameters (for example LNA: return loss, isolation) are set to reasonable values. They also influence the simulation results. The specification of all block parameters can be found in the Section 9.2.

LNA and BB Amp

Table 10-3 and Table 10-4 show the parameters for the LNA elements used for the passband LNA and the baseband amplifiers. The gain of the passband LNA is set to 20 dB (specification 5-20 dB), input impedance, noise figure and IIP3 are set according to the specification.

Table 10-3. LNA parameters

Parameter	Value
Noise Figure (dB)	2.5
Input referred IP3 (dBm)	-5
Gain (dB)	20
Reverse isolation (dB)	100
Reference impedance of port 1 (Ω)	50
Reference impedance of port 2 (Ω)	50
Input return loss (dB)	-100
Output return loss (dB)	-100

Other parameters defined in the specification (for example 1dB CP) are not part of the LNA model. The gains of the baseband amplifiers are set to 54 dB (specification 12-62 dB) which results in a receiver output power of 4 dBm. The input impedance and noise figure are also set according to the specification.

Table 10-4. BB amp parameters

Parameter	Value
Noise Figure (dB)	5
Input referred IP3 (dBm)	-10
Gain (dB)	54
Reverse isolation (dB)	100
Reference impedance of port 1 (Ω)	50
Reference impedance of port 2 (Ω)	50
Input return loss (dB)	-100
Output return loss (dB)	-100

Mixer

Table 10-5 shows the parameters of the mixer blocks. Input impedance, gain, noise figure, IIP2 and IIP3 are set according to the specification. All mixers work at the same LO frequency of 2.6 GHz. Unspecified parameters of the model are the LO power and *isolation* values which are not itemized in the table. The declaration of LO power is used to exclude the impact of LO power on the gain of the mixer. The isolation values determine the strength of signal transmission from one port to another one. For example the transmission of phase noise from the oscillator to the first mixer output can be influenced with *LO to OUT isolation*.

Table 10-5. Mixer parameters

Parameter	Value
Gain (dB)	5
Input impedance (Ω)	50
Output impedance (Ω)	25 (50 in the second mixer stage)
Input impedance for LO (Ω)	100
Input referred IP2 (dBm)	20
Input referred IP3 (dBm)	5
SSB Noise Figure (dB)	10

Lowpass filter

For the specified signal characteristics an appropriate bandpass filter model is difficult to realize. Therefore a 5th order *Chebyshev* lowpass filter is used. The properties of the lowpass filter are shown in Table 10-6. The input impedance and the corner frequency are set according to the specification.

Table 10-6. Lowpass filter parameters

Parameter	Value
Filter order	5
Input impedance (Ω)	50
Output impedance (Ω)	50
Corner frequency (Hz)	9M
Insertion loss (dB)	0

Oscillator

The parameters of the oscillator determine output frequency (2.6 GHz) and signal power (-30 dBm) as well as several noise properties. The noise parameters (for example phase noise) are used for *small-signal noise* analysis. They can be seen in Table 10-7. Phase noise is a drawback of oscillators and PLLs which must not be neglected. Phase noise modeling and simulation within transmission systems is an important task.

Table 10-7. Oscillator parameters

Parameter	Value
Output frequency (Hz)	2.6G
Output power (dBm)	-30
Output impedance (Ω)	50
Noise floor (dBc/Hz)	-150
Frequency point f1 (Hz)	2M
Phase noise at f1 (dBc/Hz)	-103
Corner frequency (Hz)	0

Phase shifter

Table 10-8 shows the properties of the phase shifter which combines a shifter and a signal splitter. The operation frequency is set to 2.6 GHz.

Table 10-8. Phase shifter parameters

Parameter	Value
Operating frequency (Hz)	2.6G
Internal resistance (Ω)	100

10.4.2 Simulation of the WLAN receiver

Frequency conversion

The depiction of the *frequency conversion* inside the WLAN receiver can be realized using *SpectreRF* [Cad03a] with a *Periodic Steady State* (PSS) analysis. This *large-signal* analysis uses harmonics of a determined beat frequency (also known as PSS fundamental frequency) to calculate the output signal. In the PSS setup the beat frequency can be calculated automatically according to the available sources and their frequencies. For a better visualization of the results a beat frequency of 50 MHz is chosen. Figure 10-17 shows the results of the PSS simulation calculating 180 harmonics. The RF signal lies at 5.2 GHz with a power amplitude of -76.5 dBm. The LO signal frequency is 2.6 GHz with a chosen amplitude of -30 dBm. As already mentioned the impact of the LO power on the receiver gain is neglected. The third part of the figure shows the signal at the first mixer output, where the RF signal is mixed down to 2.6 GHz and amplified by 24 dB. Another mixing product lies at 7.8 GHz but remains unconsidered. The inphase output with an amplitude of about 4 dBm is shown last. This DC signal is not filtered because only a lowpass filter is used in the receiver model.

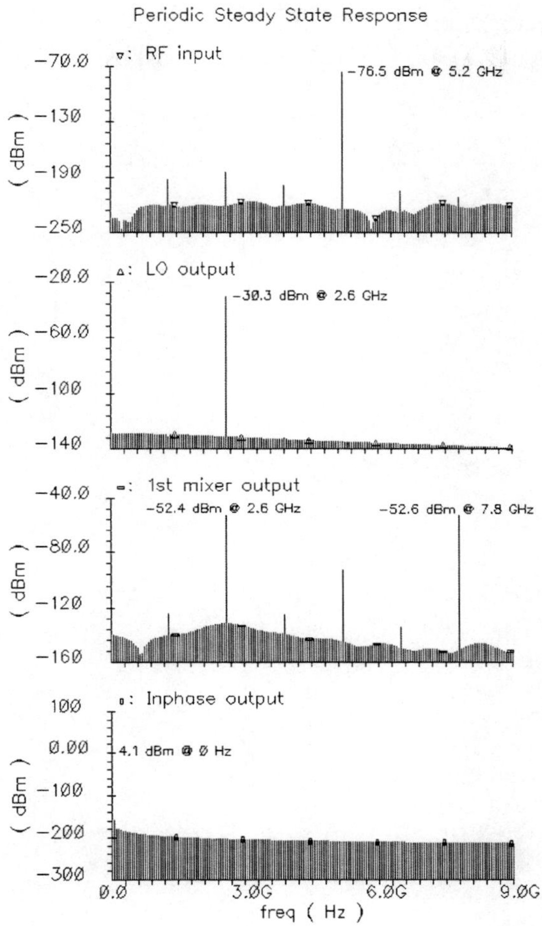

Figure 10-17. Frequency conversion of the receiver

Conversion gain

The *conversion gain* can be simulated using PSS and a subsequent *Periodic Transfer Function* (PXF) analysis. PXF is a small-signal analysis which analyzes the frequency conversion over the whole receiver. The signal contribution of all the inputs to one output is calculated. The beat frequency for PSS and PXF analysis is set to 2.6 GHz according to RF = DC and LO = 2.6 GHz input signals. In the PXF settings the frequency sweep range is set from 1 Hz to 300 MHz. The number of *sidebands* is set to three to generate four curves at multiples of 2.6 GHz. As output voltage the net *I_out* (inphase output) is chosen. Figure 10-18 first shows the contributions of all sidebands to *I_out* and then the contribution of only the input signal (sidebands -2 and 2). The conversion gain of the receiver, from input to the inphase output, can be read off as approximately 80 dB. The defined *3dB corner frequency* of the lowpass filter (9 MHz) is situated at 5.209 MHz.

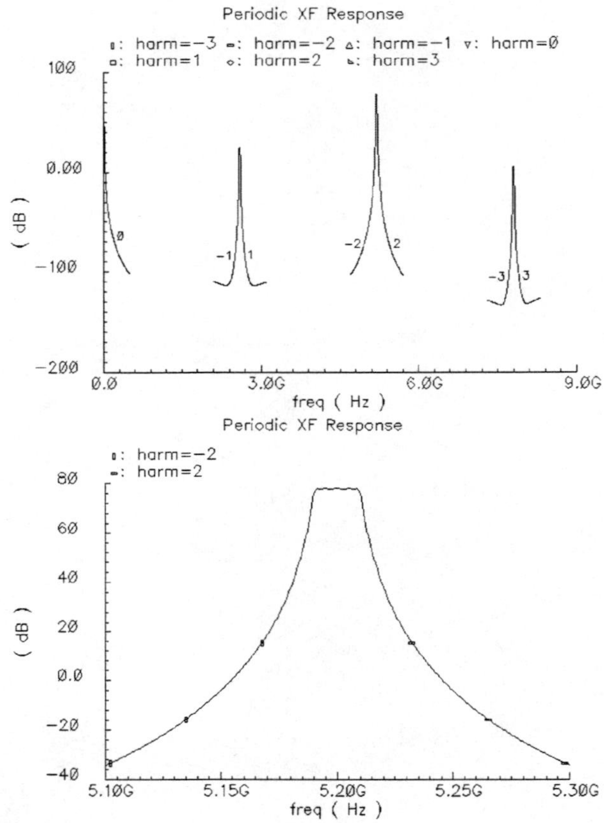

Figure 10-18. Conversion gain

Noise figure

The *noise figure* of the receiver model can be simulated using a PSS and a small-signal *PNoise* analysis. Figure 10-19 shows the overall noise figure at *I_out* with a measured value of 7.2 dB. The PSS settings are the same as the conversion gain measurement using PSS/PXF analysis. In the PNoise settings the frequency sweeps from -100 MHz to 100 MHz, which is a reasonable area at the baseband output. The number of maximum sidebands is set to 10 to ensure that enough sidebands contribute noise to the output. The positive output node is set to *I_out* and the source is set to the RF input port. The reference sideband is $k = 2$ according to |fin| = |fout + k · fpss|, where fin = 5.1 GHz … 5.3 GHz, fout = -100 MHz … 100 MHz and fpss = 2.6 GHz.

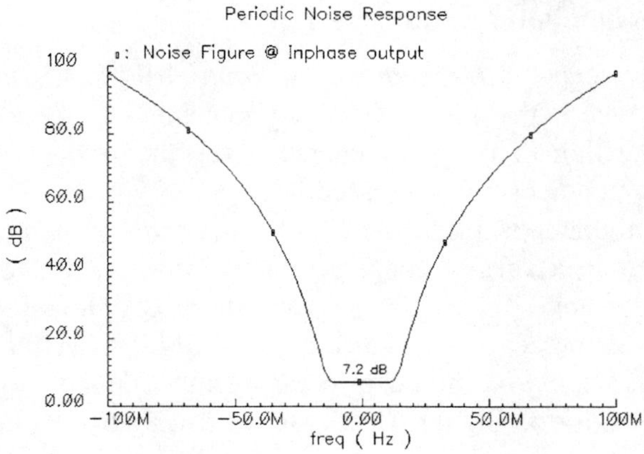

Figure 10-19. Noise figure

Phase noise

An important contribution to the noise of a system is the *phase noise* which can be seen in Figure 10-20. An oscillator does not produce a signal that runs exactly at one frequency. Small variations in the zero-crossings of the signal (called *jitter* in the time domain) result in phase noise. A further but marginal noise contribution of oscillators is the *amplitude noise*, which indicates small variations in the signal amplitude.

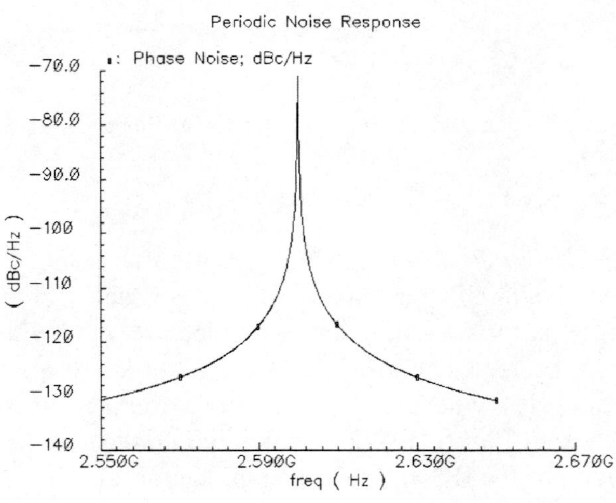

Figure 10-20. Phase noise

1dB Compression Point (input referred)

The input referred *1dB Compression Point* (1dB CP) is the value of the input power where the gain of the component is 1dB below the ideal 1dB/1dB projection. This point characterizes the region of input power where the output power is compressed.

The measurement is performed by a *swept PSS*, which means a PSS analysis sweeping a variable or a parameter, in this case the input power over a defined range. The input power range used for simulation of the WLAN receiver spreads from -80 dBm up to -10 dBm. After simulation the CP is calculated applying the postprocessing function *Compression Point* in the PSS results form. Figure 10-21 shows the 1dB CP measured at the inphase output.

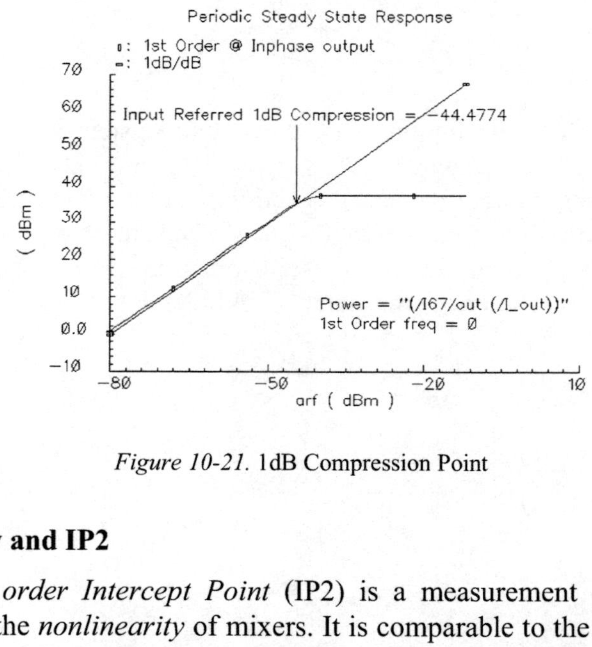

Figure 10-21. 1dB Compression Point

Nonlinearity and IP2

The *2nd order Intercept Point* (IP2) is a measurement often used to characterize the *nonlinearity* of mixers. It is comparable to the IP3, the main difference can be seen in the frequencies which are applied to construct the intercept point. The used 3rd order frequencies of the IP3 are adjacent to the 1st order frequencies whereas the 2nd order frequencies of the IP2 lie outside of the band. Figure 10-22 clarifies the context at the first LNA. A closer description of the IP3 can be found in Section 11.2.

IP2 is the value of input power where the extrapolated 2nd order sideband crosses the extrapolated output signal of the fundamental (1st order) frequency. Two signals f1 and f2 are used to cause *intermodulation distortion* and to create intermodulation products. For an IP2 measurement one of the 2nd order intermodulation products f1 - f2 or f1 + f2 is necessary.

IP2 measurement can be performed using a combined swept PSS/PAC (*Periodic AC*) or only a swept PSS analysis.

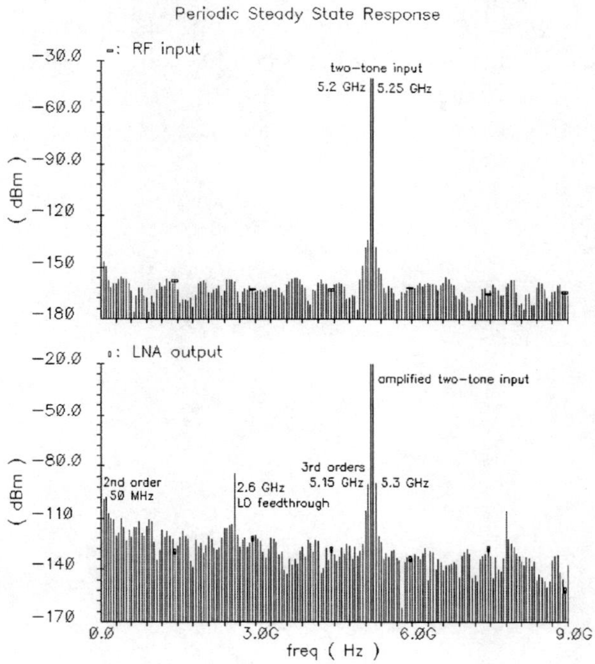

Figure 10-22. Distortion due to nonlinearity

For an exemplary measurement of IP2 a single mixer model (*rfLib*) is used. The model and the corresponding signal sources and sinks are depicted in Figure 10-23.

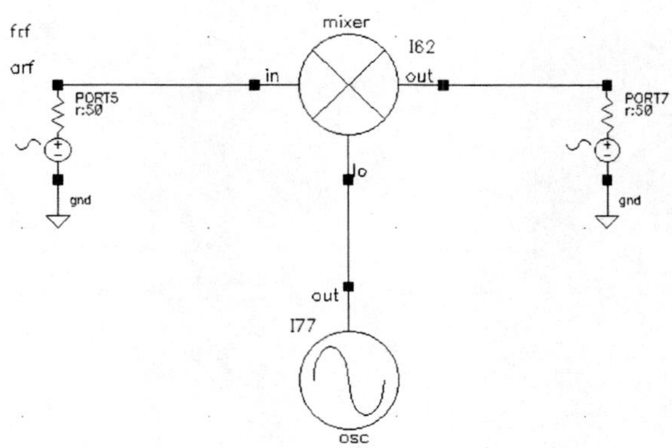

Figure 10-23. Mixer test-bench

Figure 10-24 shows the simulation results of the IP2 measurement. The input power is swept over a defined range between -60 dBm and 40 dBm. The extrapolated 2nd order harmonic crosses the extrapolated 1st order harmonic at an input power level of 19.9 dBm according to the specified parameter of 20 dBm. This intersection point is called input referred IP2.

The used input frequencies are fl = 5.2 GHz flo = 2 GHz. For the measurement the corresponding frequency pair fl - flo and 2 · (fl - flo) is chosen.

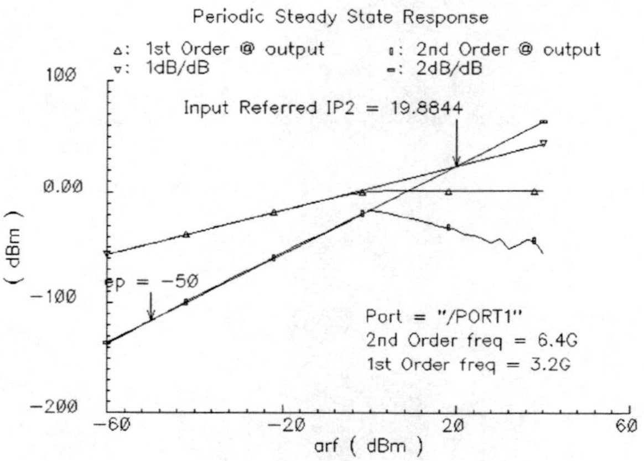

Figure 10-24. IP2 of the mixer

Chapter 11

CHARACTERIZATION FOR BOTTOM-UP VERIFICATION

11.1 Concept of Characterization

Characterization can be considered as the manual or automated determination of characteristics and parameters of a component. A component which is characterized is referred to as a *Design Under Test* (DUT). In the case of characterization by simulation it is referred to as a DUT model. Within a design flow a component may be a subsystem (for example mixer or LNA) of the communication system. Characterization can be done from:

- Measurement of manufactured components
- Simulation of component models (for example circuit level descriptions)

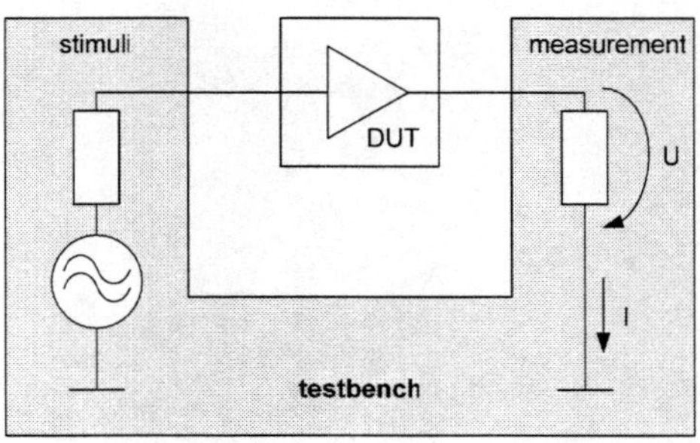

Figure 11-1. DUT within a testbench

For a characterization the DUT has to be inserted in a testbench (Figure 11-1). The testbench provides the stimuli (analog or digital signal sources) to the input ports and the measurements of the DUT output signals. For a complete characterization of a DUT a set of measurements with different stimuli may be necessary. Additional postprocessing is performed to compute the parameters of the DUT from the measured data.

Characterization based on simulation is described in the following sections. Examples show the development of testbenches and control scripts which allow the automation of simulation and postprocessing.

11.2 RF Characteristics and Parameters

Characteristics and parameters are extracted to document components and to parameterize behavioral models. Additionally, characteristics can be used for the generation of behavioral models which is explained shortly in Section 11.3. One main difference between parameters (for example 1dB CP) and characteristics (for example *AM/AM conversion*) is the fact that parameters are often parts of mathematical relationships whereas characteristics may represent a mathematical relationship. Depending on its step size a characteristic provides higher accuracy during representation of circuit properties. Characteristics can be stored in tables and may be multidimensional. Examples of important characteristics are depicted below.

The AM/AM conversion represents the dependency of the output signal power on the input signal power. The gain can be read off in the linear area of the curve. In the area of saturation the output power is compressed and the gain therefore decreases. The AM/AM conversion curve is the base for the measurement of the 1dB CP (Figure 11-2). For the extraction of this curve a periodic steady state (PSS) analysis can be used, where the input power is swept.

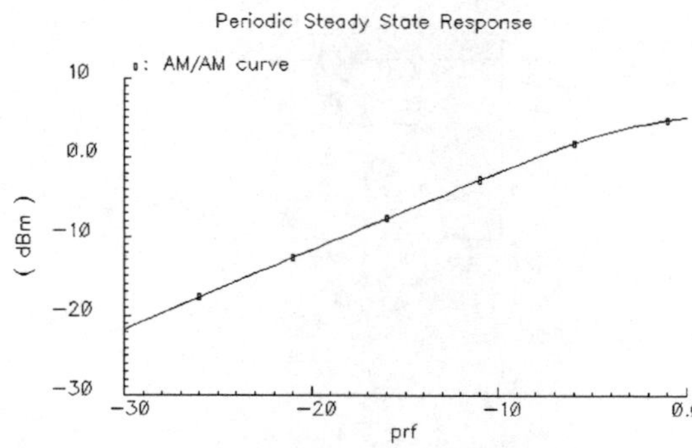

Figure 11-2. AM/AM conversion

In the saturation area of RF components the phase of the input signal is often shifted. Therefore another characteristic arises, the *AM/PM conversion* (Figure 11-3). The output signal phase depends on the input signal power. In addition to the already mentioned conversion types there also exist PM/AM and PM/PM conversions. They only play a secondary role, PM/PM conversion for example can be encountered at mixers or IQ-modulators.

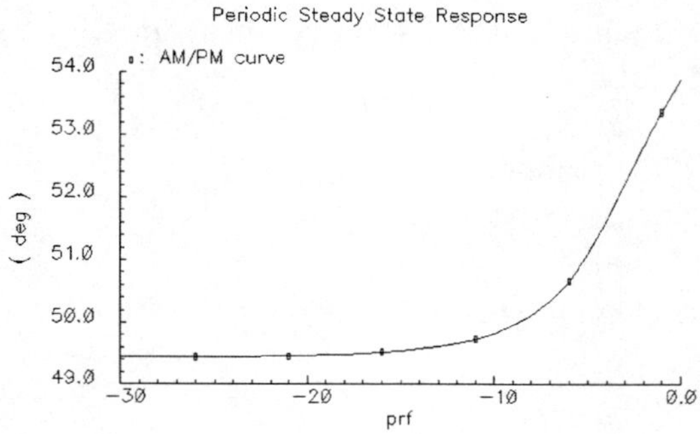

Figure 11-3. AM/PM conversion

The aforementioned aspect of multidimensional characteristics lead to characteristics which depend on more than one input value. Apart from the dependency on the input signal power, an AM/AM conversion may for example depend on the input frequency or other signal parameters. Figure 11-4 shows AM/AM conversion curves which depend on the input frequency.

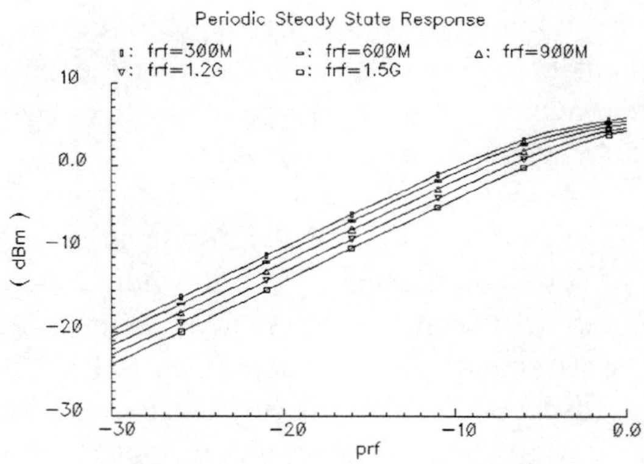

Figure 11-4. AM/AM conversion depending on frequency

In comparison to characteristics, parameters can be chosen with respect to the following points:

- Parameters that are needed to document the characteristics of a component
- Parameters that must be calibrated or optimized in a behavioral model

The parameters which are needed for documentation may differ from parameters of the behavioral model.

The most important parameters to document for RF components may be categorized in the following way:

- Impedances at the component ports
- Frequency response, S-parameters
- Gain or attenuation
- Nonlinearity (intercept points, compression points)
- Noise

Some parameters are described below.

S-parameters can be used to represent impedances and frequency response in the small-signal region. The port impedances are important parameters to ensure correct matching to the component environment. They may depend on the signal frequency.

The *frequency response* is significant for filters. Parameters include for example *corner frequencies* for lowpass and highpass filters, and *center frequency* and *bandwidth* for bandpass and bandstop filters. Other components like amplifiers or mixers may also have a frequency response because of the limited transit frequency of their active devices.

Each component of an RF system influences the power level. Active components often amplify the signal (gain), while passive components (filter, power splitter) come with an attenuation.

The measurement of parameters of the nonlinearity is important for the characterization of active components. Due to saturation effects the gain of amplifiers and mixers will decrease with increasing input level. This effect is represented by the parameter *Compression Point* (CP). Other nonlinearities produce harmonic distortions. Their strength can be depicted by *Intercept Points* (IP). The determination of compression points and intercept points is explained in more detaile in the following section.

Figure 11-5 shows a plot of the 1dB compression point (1dB CP). The output power of a mixer is plotted versus its input power level. The plot was generated from a PSS simulation with an input power sweep. A straight line is drawn 1dB below the linear area of the mixer (in the example the linear gain is -3.4 dB). The 1dB CP is the point where the output power crosses this line. Its value can be read off from the x-axis (input power = input referred) or from the y-axis (output power = output referred).

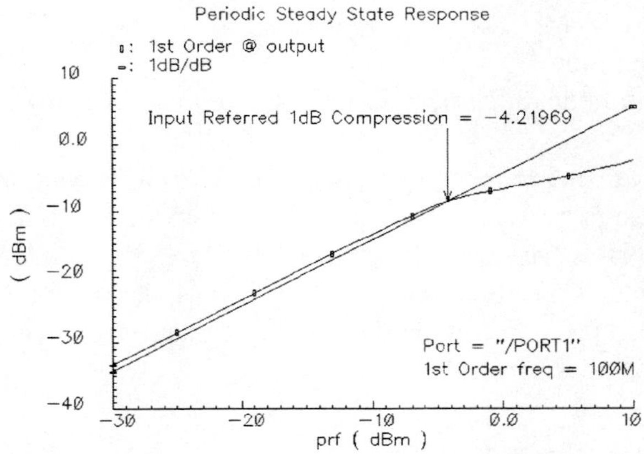

Figure 11-5. 1dB compression point

A two-tone input signal is used for the determination of the 3rd order intercept point (IP3) of the mixer. For the example (Figure 11-6) the input frequencies f1 = 900 MHz, f2 = 920 MHz, and flo = 1 GHz are chosen with a power sweep from -30 to 20 dBm. The 1st order output power (flo - f1 = 100 MHz) and the 3rd order output power (flo - (2 · f2 - f1) = 60 MHz) are plotted versus the input power. Straight lines are drawn through the linear area of the curves. The intercept point of the lines marks the IP3. For the computation of the IP2 (2nd order intercept point) a 2nd order spectral line is used instead of the 3rd order line.

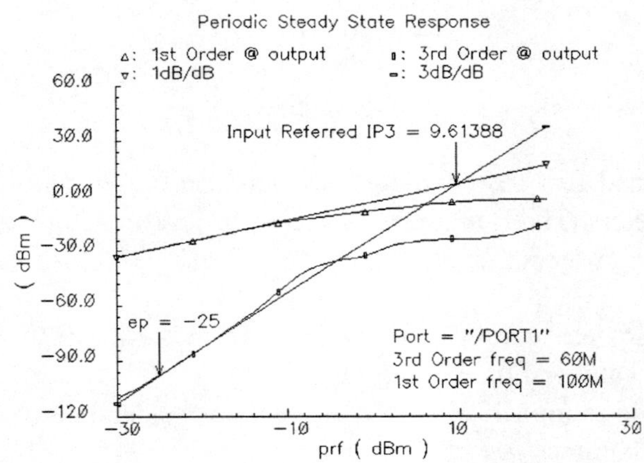

Figure 11-6. 3rd order intercept point (IP3)

11.3 Application of Characterization

Three main applications of characterization are introduced in this section:

- Model refinement
- Model generation
- Component/model documentation (basis for design reuse)

Circuit level implementations of system components are developed during the design process. They should be verified within their system environment. However, the simulation of large system parts requires efficient models. Therefore behavioral models of the components are used. Characterization provides the data necessary to configure behavioral models as accurately as possible.

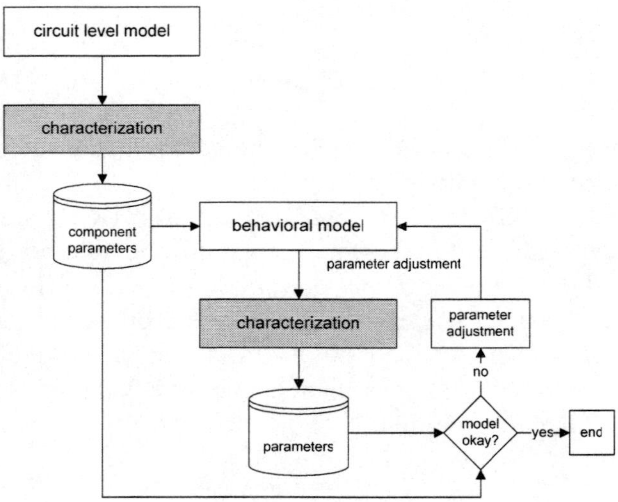

Figure 11-7. Model refinement

It is assumed that the behavioral models can be configured by a set of model parameters. The *refinement* of the behavioral parameters can be done in the following way (shown in Figure 11-7):

1. Run a characterization of the circuit level description to determine the parameter values of the behavioral model.
2. The behavioral model is inserted into a testbench and configured with the computed parameters.

3. A second characterization of the behavioral model verifies if the model accurately represents the characteristics of the circuit.
4. If the model accuracy is not good enough, the model or its parameters can be improved and again verified with the characterization.

The use of a characterization environment simplifies the configuration and verification of the behavioral model.

The *generation* of behavioral models based on the characterization of circuit level models is an efficient solution to support the bottom-up verification. The aim is to obtain a behavioral model which as close as possible represents the circuit level model. According to the simulation settings (for example step size, sweep range) the extracted characteristics, and consequently the generated model, can provide a high level of accuracy. The designer does not require knowledge about special modeling techniques, like complex baseband modeling. Figure 11-8 shows possible steps of the model generation.

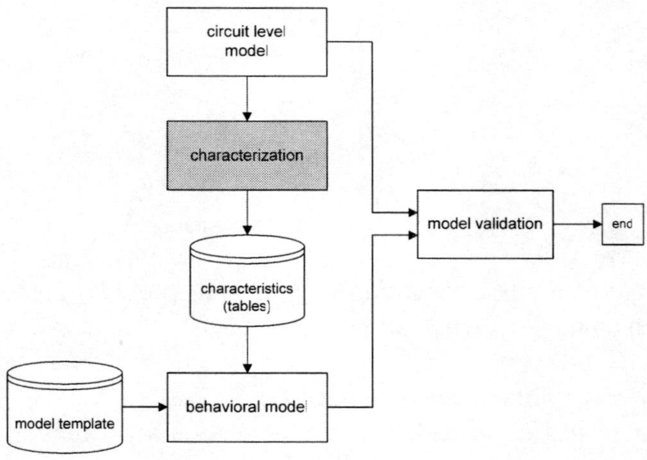

Figure 11-8. Model generation

1. A characterization of the circuit level model is performed. Several signal parameters can be swept to achieve multidimensional characteristics.
2. The characteristics can be stored in tables.
3. Each circuit type (for example LNA) requires a model template, which is implemented in the environment of the model generator.
4. The behavioral model consists of the model template and the inserted characteristics.

5. At last the circuit level model and the generated behavioral model can be compared for validation purposes (for example regarding accuracy).

Model reuse has a great importance to increase the design efficiency. A good model documentation is the base of reuse (Figure 11-9). The designer *A* has developed a circuit or a model. It is characterized to determine the information that is needed for documentation. The documentation is inserted into an intranet or internet database from which it can be found by another designer *B* using selection and search functionality.

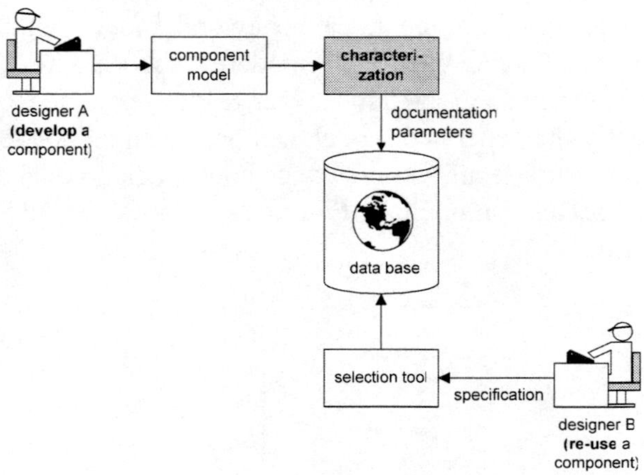

Figure 11-9. Model documentation

The usage of a characterization environment provides some advantages compared to manual documentation:

- Reduced time of the documentation
- Using standardized testbenches and simulation algorithms for parameter determination
- The format of documentation can be standardized to simplify the import into the database

11.4 Example Characterization of an LNA

An LNA model (*lnaSimple*) at circuit level is characterized to determine the parameters for a behavioral model. The schematic of the LNA is shown in Figure 11-10.

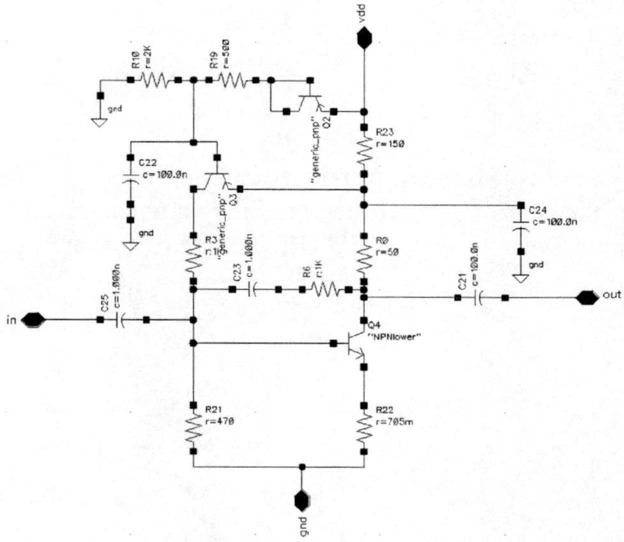

Figure 11-10. LNA schematic

The schematic of the LNA is packed into a symbol and inserted into a testbench (Figure 11-11). It consists of two port models *psin* and a *vdc* source. The left side port generates the input signal. The right side port is used to terminate the LNA and to measure the output signal.

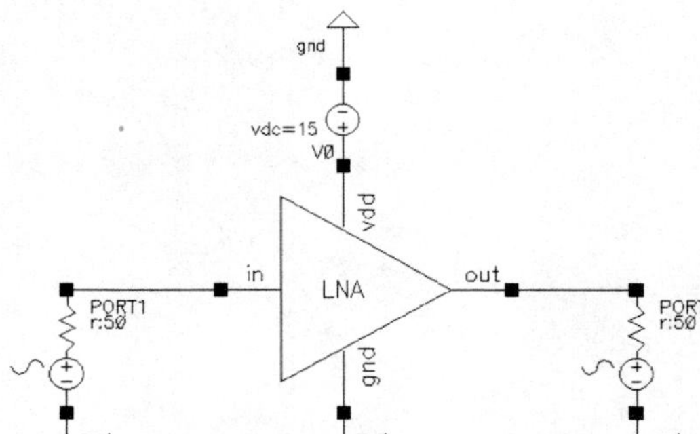

Figure 11-11. LNA testbench

The behavioral model *LNA_PB*, which is the target model supports the following parameters to be characterized:

- Available power gain
- Input and output resistance
- Input referred IP3
- Noise figure

The simulator *SpectreRF* provides the analyses and postprocessing functions for the determination of the required parameters. A PSS analysis is used to compute *power gain* and *IP3*. The resulting plots are depicted below.

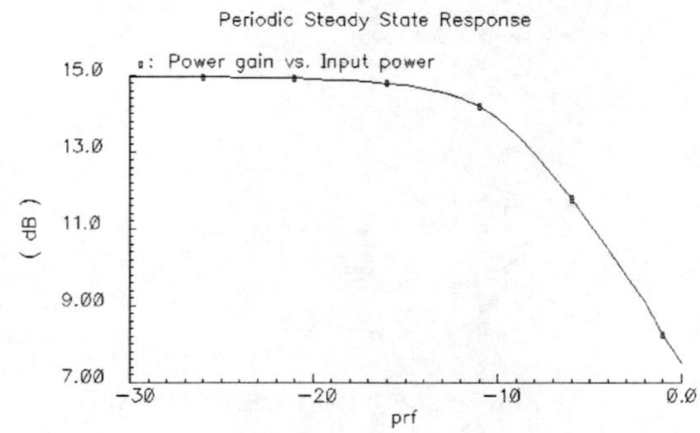

Figure 11-12. Power gain

A *power gain* of 14.9 dB is measured in the linear area of the LNA (Figure 11-12). The gain decreases with increasing input level. The 1dB compression point is reached at -10.5 dBm input power.

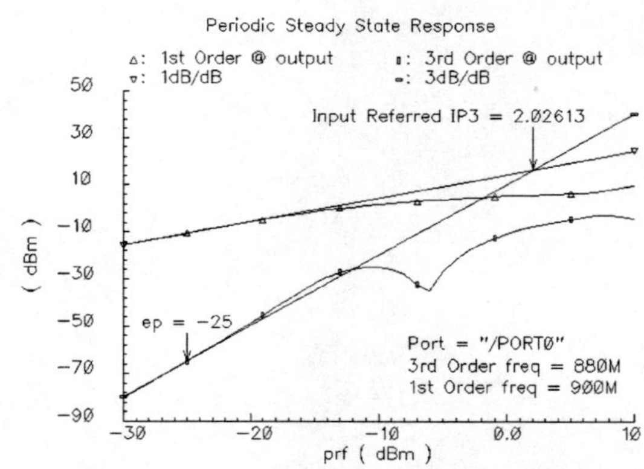

Figure 11-13. IP3

The IP3 plot (Figure 11-13) is generated using the *IPN Curves* plot function. The LNA is designed for a frequency of 900 MHz. For the extraction a two-tone signal with the frequencies 880 MHz and 900 MHz is used.

The *noise figure* is computed by means of a *PSS/PNoise* analysis. The sweep parameter can be either input frequency or input power. Figure 11-14 shows the noise figure versus the input frequency. At the operating frequency of 900 MHz a noise figure of 4.1 dB is measured.

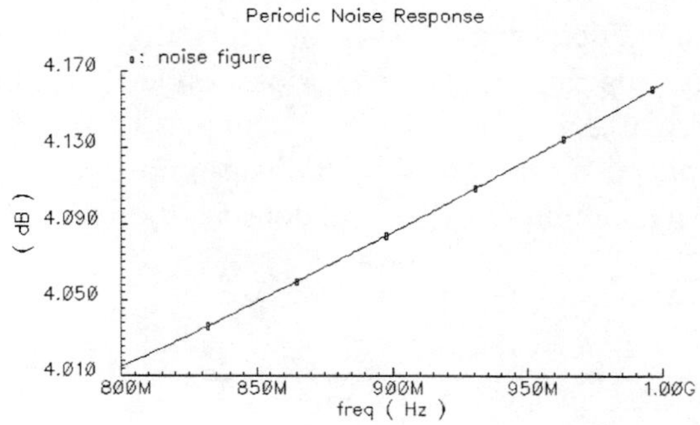

Figure 11-14. Noise figure

The *S-parameter* analysis is used for the simulation of the input and output impedances. Figure 11-15 depicts Z-parameters plotted against the input frequency. Z11 represents the input impedance and Z22 the output impedance. The impedances are complex values. An input impedance of 40.5 Ω and an output impedance of 33.7 Ω are computed at the frequency of 900 MHz.

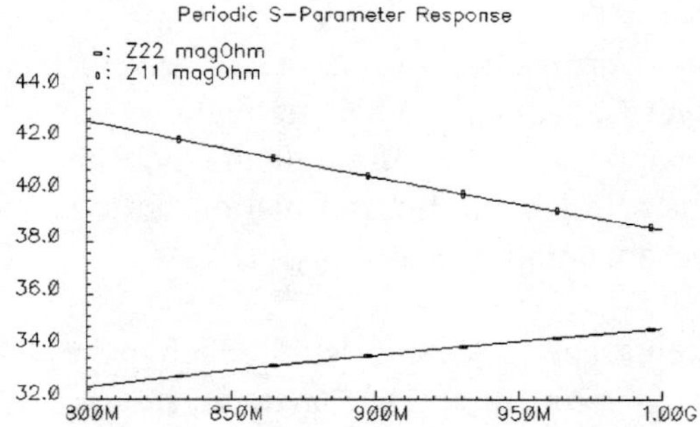

Figure 11-15. S-Parameter

The behavioral model *LNA_PB* can now be configured with the determined parameters. They are depicted in Table 11-1.

Table 11-1. LNA parameters

Parameter	Value
Gain	14,9 dB
IP3	2 dBm
Noise figure	4,1 dB
Z11	40,5 Ω
Z22	33,7 Ω

The characterization may now be repeated to verify *LNA_PB*. Differences may appear because behavioral models cannot represent all effects of a circuit level model. An example is the plot of the port impedances depicted in Figure 11-16. The impedance values are now independent from the input frequency in contrast to the circuit level model.

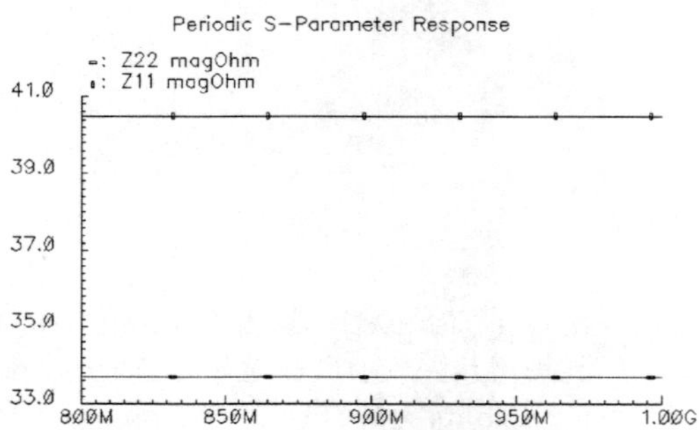

Figure 11-16. S-Parameter

11.5 Characterization Environment

An exemplary *characterization* is described in Section 11.4. It is recognizable that manual characterization is time-consuming. Characterization can be simplified through an environment which automates all steps of the characterization, like simulation settings and postprocessing. It consists of two main components:

- Testbenches represent the test circuit which provides stimuli and the measurement environment for the characterization of the DUT model. More than one testbench may be needed to characterize all required parameters of the model.
- Simulation control scripts are used to set up the analyses and to provide additional computation of parameters in a postprocessing step.

The combination of predefined testbenches and simulation control scripts simplifies the execution of a characterization. This is demonstrated in the following by using the *Cadence Design Framework II* (DFII). Figure 11-17 demonstrates the usage of the Cadence tools in a characterization environment.

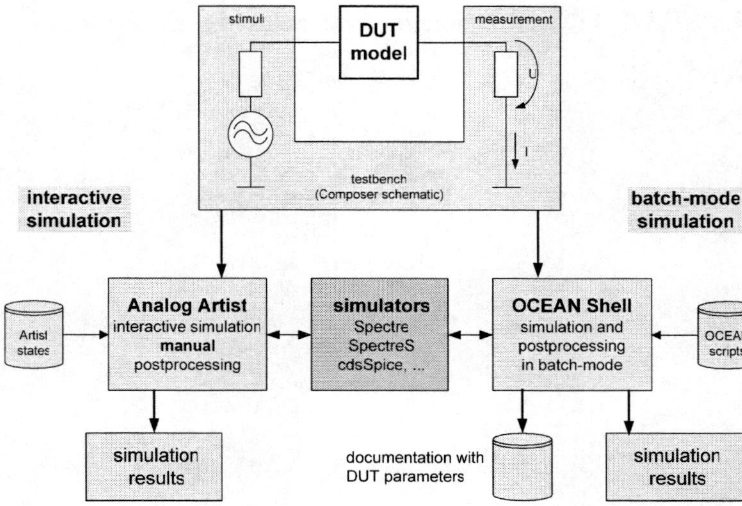

Figure 11-17. Characterization in the DFII

First, the DUT model is inserted into the testbench schematic. The schematic contains the test environment (sources and measurement blocks). The port model, *psin* for example, can be used to realize a signal source for various analyses as well as for output termination and power measurement. Predefined testbenches can be stored in libraries.

The schematic is ready for the simulation with the analog simulators after insertion of the DUT model. Then it is necessary to set up the analyses and to adjust the simulation environment settings. These configurations are also part of the characterization environment. Cadence provides two facilities to store such analysis setups. The first one is to store the analysis and plot settings of the *Cadence Analog Design Environment* (ADE) in *states*. The second facility insists on the application of *OCEAN* and *SKILL* scripts which can be used for batch mode simulation. The next section provides a short overview of the use of OCEAN scripts followed by an example.

Using OCEAN scripts for characterization

In addition to the interactive simulation with ADE it is possible to run analyses, visualization and postprocessing in a batch mode. Therefore the script languages SKILL and OCEAN are used.

- SKILL is a script language with various functions which allow to configure and control the complete DFII.
- OCEAN [Cad03b] is a subset of SKILL with functions that are needed to control the simulation environment with the appropriate postprocessing.
- Both OCEAN and SKILL functions can be used in OCEAN scripts.

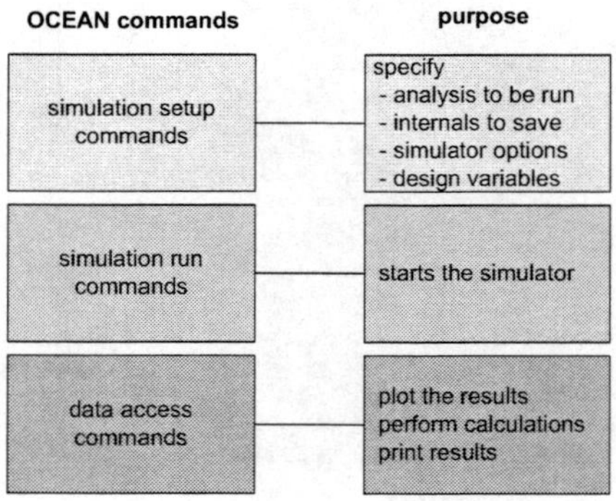

Figure 11-18. Subdivision of OCEAN commands

An easy way to develop an OCEAN script starts with an interactive simulation in the ADE. The menu command *Session->Save Script* creates an OCEAN script that contains the actual simulation settings. It can be modified and extended, for example, by several postprocessing functionalities.

The OCEAN script can be loaded in an OCEAN shell or the Cadence *Command Interpreter Window* (CIW) to run the analyses and the postprocessing. The advantages are:

- Different analyses can run subsequently
- Postprocessing and computation of parameters and characteristics is automated
- Results can be stored in ASCII files or plots

Testbench and OCEAN scripts provide an automated characterization.

Example OCEAN script of an AC analysis

A simple OCEAN script is represented in the following. It starts an AC analysis of the test circuit *lowpass*. The frequency response is displayed graphically, the insertion loss and the corner frequency of the lowpass are computed.

The simulator, the netlist file, and the output directory are specified in the first section of the script.

```
;Simulation environment for Spectre
simulator('spectre)
design("./simulation/lowpass/spectre/schematic/netlist/netlist")
resultsDir("./simulation/lowpass/spectre/schematic")
```

The AC analysis is set to the frequency range from 500 MHz to 1.5 GHz (logarithmically 100 points per decade). Operating voltage, bias current, and temperature are specified and the simulation is started.

```
; AC ANALYSIS
analysis('ac ?start "500M"  ?stop "1.5G"  ?dec "100")
desVar("UB" 2.8)
desVar("Ibias" 1.0m)
temp(25.0)
run()
```

The output directory is automatically opened after the analysis has finished. The results of the AC analysis are selected. A labeled plot window is opened and the output voltage is logarithmically plotted.

```
; Plot the resuls
acwave = selectResult("ac")
winAC=newWindow()
plot(db20( v("/OUT") ))
label = addWindowLabel(list( 0.50 0.95 ) "Frequenzgang")
```

Finally the corner frequency and insertion loss are determined from the frequency characteristic. In addition the type of filter (in the example *low* for *lowpass*) must be specified. "3.0" indicates that the 3dB corner frequency must be determined. The maximum of the difference between output and input signal is computed for the determination of the insertion loss. Insertion loss and corner frequency are finally printed in the OCEAN shell.

```
;bandwidth and insertion loss calculation
b=bandwidth(v("/OUT") 3.0 "low")
il=ymax(db20(v("/OUT")) - dB20(v("/IN")))

;print the results to the shell
printf("corner frequency: %5.3f   insertion loss: %5.3f dB\n" b il)
```

To meet today's requirements of a characterization environment special tools have been developed. They contribute especially to the bottom-up verification and provide solutions according to automated characterization, model generation and optimization. *Cadence Virtuoso Specification-driven Environment* (VSdE) has to be mentioned in this aspect.

11.6 Characterization Using the OCEAN Script Language

Objective

Designing an environment for the characterization of an LNA. The following parameters shall be determined in the frequency range of approximately 900 MHz:

- Gain
- IP3
- Noise figure

The characterization environment shall consist of a testbench schematic and an OCEAN script. It shall be tested with characterization of the behavioral model *LNA_PB* from *rfLib*. The behavioral model is predefined with gain = 20 dB, IP3 = -10 dBm, noise figure = 2 dB and impedances each of 50 Ω.

Proposed solution

The following three sections describe how to design a characterization environment based on the OCEAN script language.

11.6.1 Creation of the testbench schematic

As already described, a characterization environment basically consists of a testbench which embeds the DUT, a source which provides the stimuli, a sink, and several analysis and postprocessing settings.

The testbench of the characterization environment must ensure that all analyses, for example small- and large-signal, can be used in one characterization flow without changing the testbench manually in between. Small-signal analyses in *SpectreRF* like *PNoise* require that the source operates as a DC source. Since the signal types of a source cannot be controlled by design variables a switch must be used. A switch connects the DUT with different ports. It has to be mentioned that the original switch *sp2tswitch* (*analogLib*) must be modified before using in your own testbench. A Verilog-A module which switches between the different sources could also be used as an alternative.

Besides the signal types both sources provide the design variables *prf* (power amplitude in dBm), *frf1* (first signal frequency) and *frf2* (second signal frequency, only in sine source). The switch is controlled by a fourth design variable called *switch_pos*. For the sink a DC port is used. Figure 11-19 shows the complete testbench with the inserted DUT and its parameters.

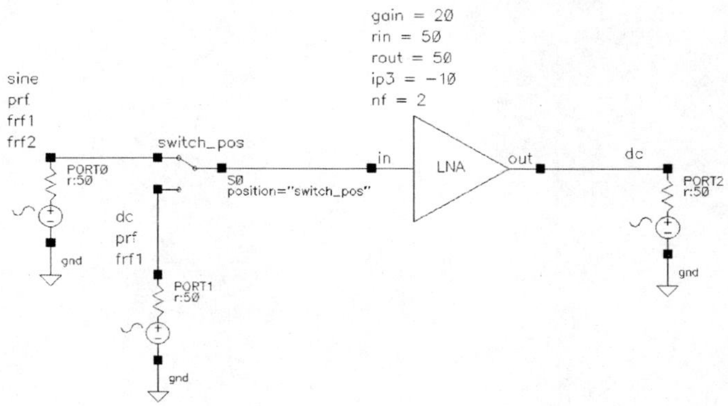

Figure 11-19. Testbench schematic

11.6.2 Analysis settings and simulation

The analysis types to use for the extraction of the specified parameters can now be considered. The following combinations are proposed.

- Gain — PSS analysis
- IP3 — Two-tone PSS analysis with sweeping input power
- Noise figure — PSS/PNoise analysis

The design variables for the gain measurement are *prf* = -40 dBm, *frf1* = 900 MHz, and *frf2* = 0. The variable *switch_pos* is set to 1 to ensure that the sinusoidal source is used for this measurement. The PSS analysis settings include a beat frequency of 900 MHz. An output harmonic of 1 is sufficient to calculate the gain at this frequency.

Now the first simulation run can be started. After simulation the gain is plotted using the function *power gain* in the PSS results form. Figure 11-20 depicts a value of approximately 20 dB. In the ADE the analysis settings can be saved in a first OCEAN script which is shown below.

```
simulator('spectre)
design("./simulation/LNA_PB_lab/spectre/schematic/netlist/netlist")
resultsDir("./simulation/LNA_PB_lab/spectre/schematic")
analysis('pss ?fund "900M"  ?harms "1"  ?errpreset "moderate")
desVar("frf2" 0)
desVar("frf1" 900M)
desVar("prf" -40)
desVar("switch_pos" 1)
save('i "/PORT0/PLUS" "/PORT1/PLUS" "/PORT2/PLUS")
```

```
temp(27)
run()
```

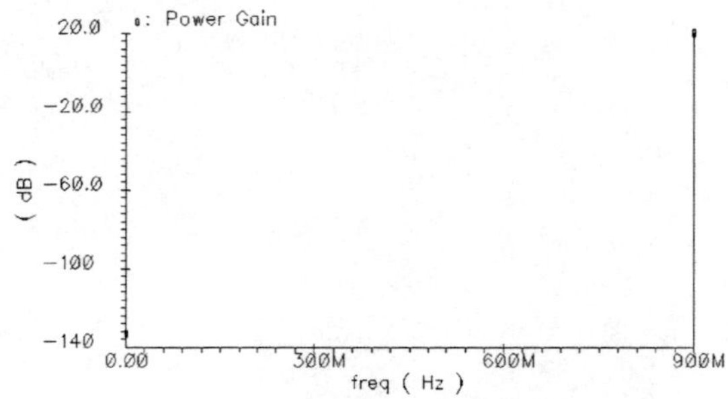

Figure 11-20. Power gain

For the *noise figure* measurement the *PSS/PNoise* analysis is applied. The design variable *switch_pos* is set to 2 to connect the DUT with the DC source. All other design variables are left unchanged. The PSS beat frequency is set to 900 MHz. The number of output harmonics is 1. In the PNoise settings the frequency sweep is set from 700 MHz to 1200 MHz. This range is a reasonable area to consider the impact of noise. The maximum count of sidebands is set to 10 and the reference sideband is 0, because no frequency conversion occurs between input and output. Output *net6* has been chosen and the input port is *PORT1*.

The noise measurement can be executed next. In the PNoise results form the noise figure can be directly plotted. It is calculated with approximately 2.08 dB (Figure 11-21) and complies with the specified model parameter. The settings can be saved in an OCEAN script.

```
simulator('spectre)
design("./simulation/LNA_PB_lab/spectre/schematic/netlist/netlist")
resultsDir("./simulation/LNA_PB_lab/spectre/schematic")
analysis('pss ?fund "900M"  ?harms "1"  ?errpreset "moderate")
analysis('pnoise ?start "700M"  ?stop "1.2G"  ?maxsideband "10" ?p
"/net6"  ?n ""  ?oprobe ""  ?iprobe "/PORT1" ?refsideband "0")
desVar("frf2" 0)
desVar("frf1" 900M)
desVar("prf" -40)
desVar("switch_pos" 2)
save('i "/PORT0/PLUS" "/PORT1/PLUS" "/PORT2/PLUS")
temp(27)
run()
```

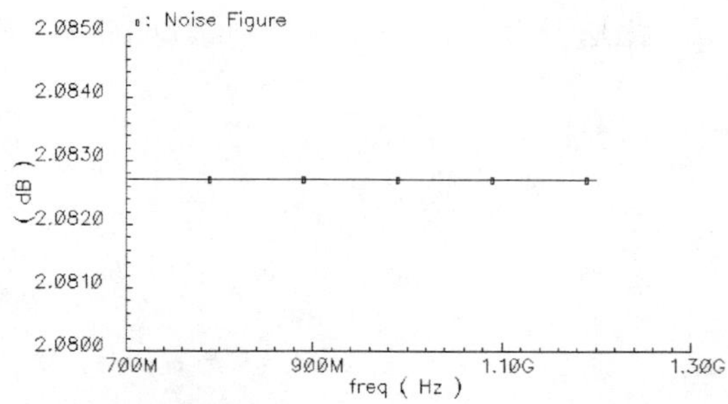

Figure 11-21. Noise figure

The *IP3* measurement requires a second sinusoidal tone. Therefore the sinusoidal source is switched to the DUT (*switch_pos* = 1) and the design variable *frf2* is set to 910 MHz, which is closely adjacent to *frf1* = 900 MHz. In the *PSS* settings the beat frequency is now automatically calculated to 10 MHz. To reach the output values of approximately 900 MHz the number of harmonics is set to 100. Therefore the largest output frequency can be 100 · 10 MHz = 1 GHz. In the sweep section of the analysis settings form the input power *prf* is swept from -50 dBm to 0 dBm. A small step size increases the accuracy and may be set to 1.

The settings for a swept PSS are complete and the simulation can now be started. For an IP3 measurement the *PSS/PAC* (PAC - Periodic AC) approach could also be used. In the results form the function *IPN curves* is used to plot the IP3. The input power is set to *variable sweep* and the *extrapolation point* can be set to -45 dBm, which lies in the lower third of the sweep range. The most important part to construct the IP3 is the choice of the correct frequencies. In this case, for example, a 1st order frequency of $fout1 = frf2 = 910$ MHz and a 3rd order frequency of $fout2 = 2 \cdot frf1 - frf2 = 890$ MHz are chosen. The value of the input referred IP3 is -10 dBm according to the specified LNA parameter (Figure 11-22). As described for the first two parameters the associated OCEAN script can be generated in the ADE.

```
simulator('spectre)
design("./simulation/LNA_PB_lab/spectre/schematic/netlist/netlist")
resultsDir("./simulation/LNA_PB_lab/spectre/schematic")
analysis('pss ?fund "10M"  ?harms "100"  ?errpreset "moderate"
?param "prf" ?start "-50" ?stop "0" ?step "1")
desVar("frf2" 910M)
desVar("frf1" 900M)
desVar("prf" -40)
desVar("switch_pos" 1)
save('i "/PORT0/PLUS" "/PORT1/PLUS" "/PORT2/PLUS")
temp(27)
run()
```

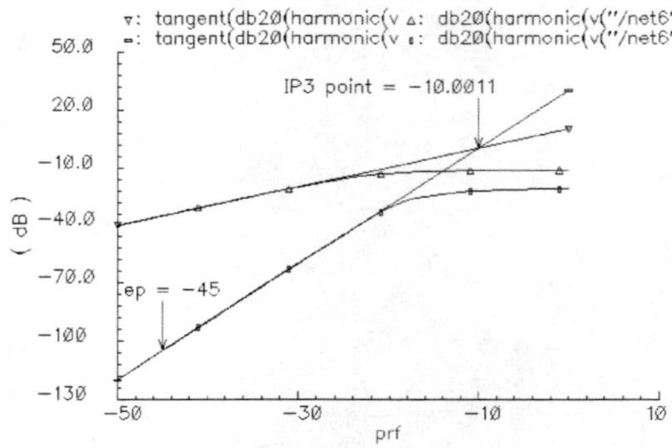

Figure 11-22. IP3

As can be seen, all parameters which were defined in the LNA model are correctly extracted. Therefore the testbench and the analysis settings are suitable for a simple characterization environment.

11.6.3 Combination and extension of the OCEAN scripts

The three generated OCEAN scripts follow the same scheme. The design environment settings (for example simulator, design, results directory) are located at the top. The analyses are then defined as well as the design variables. Several nodes are saved and the temperature is set. The command *run()* starts the simulation.

The scripts do not contain commands for plot and postprocessing capabilities, therefore it is necessary to manually extend them. Furthermore the scripts must be merged to realize the extraction of parameters within one characterization run.

The choice of the following OCEAN commands [Cad03b] can be used to extend the scripts:

```
selectResults     # selects the results from a particular analysis
plot              # plots waveform
value             # returns the Y value of a waveform
                    for a given X value
ip3Plot         # plots the IP3 curves
ipn               # performs a nth-order intercept measurement
harmonic          # returns the waveform for a given harmonic index
```

The first OCEAN script is extended by a section which plots the *power gain*. First the PSS frequency domain (*pss_fd*) is selected as the result. Then a window is opened to plot the power gain from the PSS data. A complex command for the calculation is necessary. The result is stored in the variable *powergain* and plotted using the *plot* command. Before the definition of the second analysis the current settings must be deleted.

```
; plots power gain
selectResults('pss_fd)
newWindow()
powergain = db10((let(((vn (v("/net6") - 0.0))) spectralPower((vn /
resultParam("PORT2:r")) vn)) / harmonic(spectralPower(-
i("/PORT0/PLUS") (v("/net08") - 0.0)) '1)))
plot( powergain ?expr '("Power Gain"))

; deletes analysis settings of first simulation run
delete('analysis)
```

In the second OCEAN script the commands to plot the *noise figure* are added. For this purpose the appropriate result (*pnoise*) must be selected. *NF* is a predefined variable which contains the plot data for the noise figure. As seen in the first script extension the analysis settings must be deleted.

```
; plot noise figure
selectResults('pnoise)
newWindow()
noisefigure = getData("NF")
plot(noisefigure ?expr '("Noise Figure"))

; deletes analysis settings of second simulation run
delete('analysis)
```

The last OCEAN script for *IP3* measurement must be extended by special intercept point plot functions. After selecting the results (*pss_fd*) a special function (*ip3Plot*) plots the curves which are necessary for IP3. It requires information about the net to examine, the sideband indices (89, 91) and the extrapolation point (-45). Furthermore, two variables are used to store the data for the 1st order frequency (*refWave*) and the 3rd order frequency (*spurWave*). The function *ipn* uses the variables to perform an IP3 measurement.

```
; plots IP3 curves and prints output for IP3 value
selectResults('pss_fd)
newWindow()
ip3Plot(v("/net6") 89 91 -45)
spurWave=dB20(harmonic(v("/net6") 89))
refWave=dB20(harmonic(v("/net6") 91))
ip3_loc=ipn(spurWave refWave 3 1 -45 -45)
```

The complete OCEAN script for the characterization of the *LNA_PB* is shown below. This simple environment can also be used to characterize other amplifiers. To obtain a good overview a few comments have been added.

Complete OCEAN script

```
; environment settings
simulator('spectre)
design("./simulation/LNA_PB_lab/spectre/schematic/netlist/netlist")
resultsDir("./simulation/LNA_PB_lab/spectre/schematic")

; analysis settings and design variables for gain measurement
analysis('pss ?fund "900M"  ?harms "1"  ?errpreset "moderate")
desVar("frf2" 0)
desVar("frf1" 900M)
desVar("prf" -40)
desVar("switch_pos" 1)
save('i "/PORT0/PLUS" "/PORT1/PLUS" "/PORT2/PLUS")
temp(27)

; simulation start
run()

; plot power gain
selectResults('pss_fd)
newWindow()
powergain = db10((let(((vn (v("/net6") - 0.0))) spectralPower((vn /
resultParam("PORT2:r")) vn)) / harmonic(spectralPower(-
i("/PORT0/PLUS") (v("/net08") - 0.0)) '1)))
plot( powergain ?expr '("Power Gain"))

; deletes analysis settings of first simulation run
delete('analysis)

; analysis settings and design variables for noise measurement
analysis('pss ?fund "900M"  ?harms "1"  ?errpreset "moderate")
analysis('pnoise ?start "700M"  ?stop "1.2G"  ?maxsideband "10" ?p
"/net6"  ?n ""  ?oprobe ""  ?iprobe "/PORT1" ?refsideband "0")
desVar("frf2" 0)
desVar("frf1" 900M)
desVar("prf" -40)
desVar("switch_pos" 2)

; simulation start
run()

; plot noise figure
selectResults('pnoise)
newWindow()
noisefigure = getData("NF")
plot(noisefigure ?expr '("Noise Figure"))

; deletes analysis settings of second simulation run
delete('analysis)

; analysis settings and design variables for IIP3 measurement
analysis('pss ?fund "10M"  ?harms "100"  ?errpreset "moderate"
?param "prf" ?start "-50" ?stop "0" ?step "1")
desVar("frf2" 910M)
desVar("frf1" 900M)
desVar("prf" -40)
desVar("switch_pos" 1)
```

```
; simulation start
run()

; plots IP3 curves and prints output for IP3 value
selectResults('pss_fd)
newWindow()
ip3Plot(v("/net6") 89 91 -45)
spurWave=dB20(harmonic(v("/net6") 89))
refWave=dB20(harmonic(v("/net6") 91))
ip3_loc=ipn(spurWave refWave 3 1 -45 -45)
```

Chapter 12

ADVANCED METHODS FOR OVERALL SYSTEM SPECIFICATION AND VALIDATION

12.1 Gap between System Level and Block Level Simulation

Current electronic systems consist of analog and digital parts in most cases. Typical analog subsystems are sensors and actuators in the automation area and analog front-ends to transmission channels in the telecommunication area. The analog functionality is connected to digital units like signal processors and controllers. The system performance depends on the accuracy of the analog components, the performance of the digital algorithms, and on the proper specification of the mixed-signal interface. The system performance can be validated and improved by overall system simulation. The simulation environment must be efficient for complex DSP algorithms and accurate for analog subsystems. Specialized simulation technologies are discussed in this section by means of a wireless communication system simulation.

Digital transmission technology is used for all current date transmission standards. DSP algorithms realized in hardware and software perform source coding, forward error correction, modulation, synchronization, and other algorithms. The complexity of DSP functionality has grown with new standards for the 3rd generation wireless systems and beyond.

At the interface to the transmission channel analog and mixed-signal components (like A/D converter, mixer and amplifier) are used to adapt the modulated data to the physical transmission channel. The quality of this RF front-end has a great impact on the performance of the communication system. Nonlinearity of analog components may cause transmission errors while interferers are present.

System level simulators like ADS Ptolemy, CoCentric System Studio, MATLAB, and SPW are used in system specification. Compiled C-coded models are often used together with event or data stream driven scheduling algorithms. These models provide high simulation performance for the analysis of complex DSP algorithms. On the other hand these simulators have no special algorithms to simulate analog and mixed-signal components.

Simplified models of the analog part within the system level simulator may be sufficient for the first estimation of how the analog components influence the system performance. However, for an accurate simulation of the analog part it may be necessary to combine analog and system level simulation. Different solutions can be used to closing the gap between system-level and analog modeling:

- File coupling of simulators (Section 12.2)
- Direct cosimulation of system level and analog simulators (Section 12.3)
- Generated black box models (Section 12.4)

12.2 File Coupling of Simulators

File coupling is the simplest way to exchange data between analog and digital design domains. It can be used if no feedback between the domains exists. File input and output is available in most simulators. The data file format can be different for time discrete system simulators and continuous time analog simulators. In this case file converters must be used. This solution can be used, for example, to provide realistic test patterns to the RF designer. In this case the system level is used to generate digital modulated signals. The RF designer uses these signals to evaluate the performance of the RF subsystem. In some cases the output of the RF design is again stored in a file for further postprocessing with system level models. This solution is more applicable to support RF design than for overall system validation.

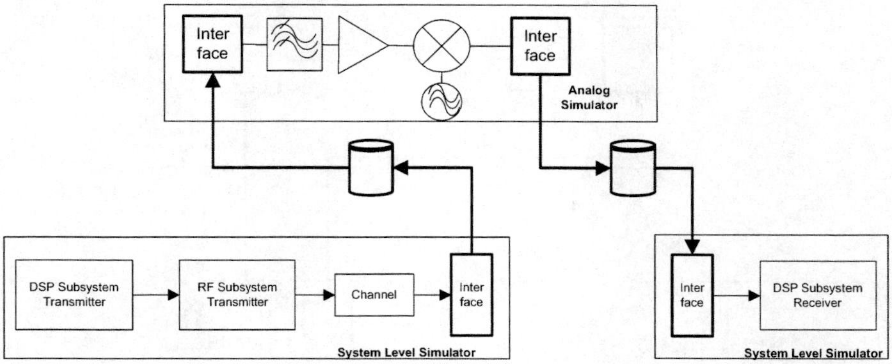

Figure 12-1. File coupling of simulators

12.3 Direct Cosimulation of System Level and Analog Simulators

In contrast to file coupling, in this solution different simulators are running simultaneously in direct cosimulation. This allows feedback loops between subsystems modeled in different simulators. The communication between the tools is usually realized by sockets or shared memory. For shared memory coupling both tools are executed on the same host. Socket connection allows the communication between tools on the same or different host in a computer network.

The implementation of a simulator coupling requires some experience in simulation and software programming. The coupling is sometimes provided by the simulator vendors. Otherwise it can be implemented by the user if both simulators have an interface for C-coded models. A C-coded interface model can then be used to exchange the data with the corresponding interface model in the other simulator. The principle of direct cosimulation is shown in Figure 12-2.

The system level simulation is a time domain analysis. It corresponds to the transient analysis of analog simulators. By coupling both simulation algorithms the time points for data exchange must be synchronized. System simulators use a discrete time scale, while analog simulators use a time continuous signal representation.

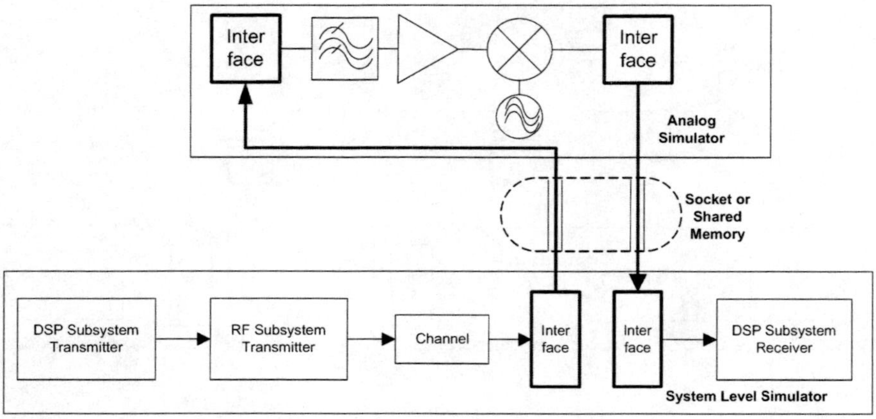

Figure 12-2. Principle of cosimulation

The main advantage of direct cosimulation is that optimized simulation tools for each system part and design level can be used. It provides the optimum accuracy level for each system part as well as debugging and visualization capabilities from both tools well suited to the design tasks. Additionally, models from both design domains can be reused in the full system simulation. In summary the direct cosimulation provides an overall system analysis with high accuracy.

Unfortunately there are some disadvantages of cosimulation. The main disadvantage is the low simulation performance as a consequence of process communication overheads and very detailed simulation of the analog parts. The development of a cosimulation interface requires some experience in simulation technology. The user needs at least some basic knowledge about each of the tools. Since both simulators are executed simultaneously the costs for software licensing is increased.

Time synchronization in cosimulation

As shown in Figure 12-3 system level simulators use a different signal representation than analog simulators. Continuous time signal representation is used in analog simulation. The width of the simulation steps varies during the simulation, depending on the gradient of the signals. The signal values are represented by a pair consisting of time and value.

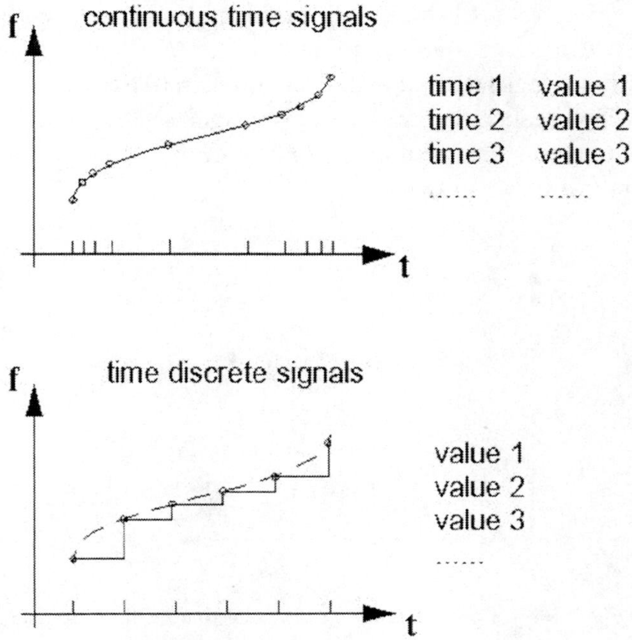

Figure 12-3. Continuous and discrete time signals

The system level simulator uses equidistant samples for signal representation. Therefore the timing information does not need to be stored with the signal values. The signal is characterized by the sequence of values and the sampling rate or sampling frequency parameter.

At the border between analog and system level domains a signal conversion must be done. This is usually realized within interface blocks. They may be generated automatically depending on the implementation of the simulator coupling.

Figure 12-4 shows the conversion from the time continuous analog domain to the time discrete signal representation. Since the discrete signal requires values at defined equidistant points, the continuous wave is re-sampled. Interpolation can be used if the sampling point is located between two computed signal values. In some cases the analog simulator can be controlled to compute additional signal points at the desired sampling times.

The opposite conversion from the time discrete domain to the continuous domain is shown in Figure 12-5. The timing information must be added into the interface blocks. The interfaces may count the samples and produce the timing value. The analog simulator updates the values from the time discrete domain only at the sampling points. Problems with convergence and performance of the analog simulation arise if the gradient between two subsequent signal values is high. This problem can be reduced by decreasing the time between successive sampling points. However this will also slow down the simulation speed in the discrete domain. Another way is the use of functions, which reduce the gradient of the signals. Such functions are often available in mixed-signal simulators. For example Verilog-AMS provides the operators *transition* and *slew*.

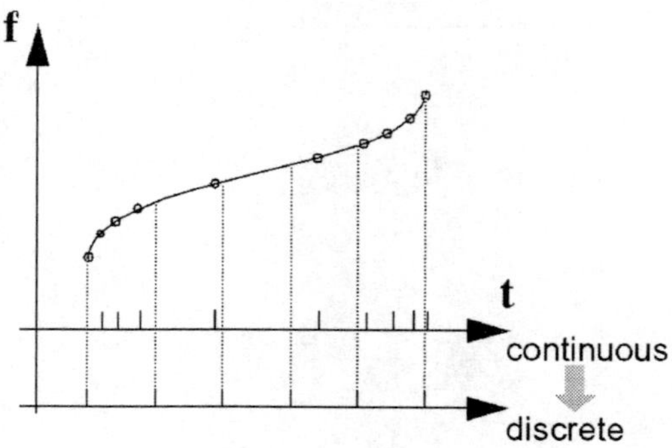

Figure 12-4. Conversion from continuous to time discrete signals

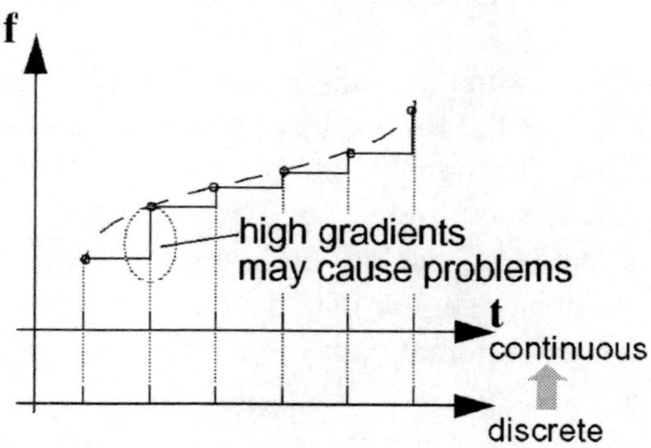

Figure 12-5. Conversion from time discrete signals to continuous signals

Cosimulation example (SPW and AMS Designer)

The usage of cosimulation is illustrated by means of the commercial cosimulation solution for the system level simulator SPW (Coware) and the mixed-signal simulator AMS Designer (Cadence). The digital baseband processing of a wireless LAN system is modeled and analyzed in SPW. A detailed analog behavioral model of the RF subsystem is included by cosimulation with AMS designer. The modeling flow is outlined below.

The RF subsystem is modeled in the Cadence Design Framework DFII. Analog behavioral models written in Verilog-A are used to describe the building blocks of the RF front-end. It contains an amplifier, filters, and a mixer. Transistor level models can also be used. Behavioral models are selected for performance reasons. The resulting model can be analyzed with the RF simulator SpectreRF as well as with the AMS Designer, which performs a transient simulation of the subsystem in the cosimulation. With the enabled AMS option, a Verilog-AMS netlist of the subsystem is generated after each "Check and Save" command.

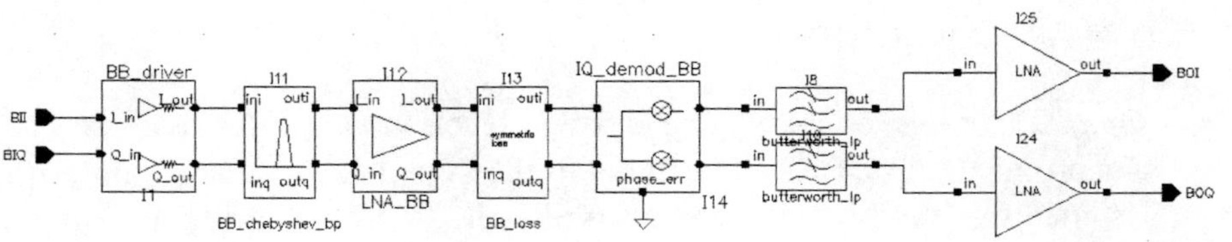

Figure 12-6. Schematic of the RF subsystem within DFII

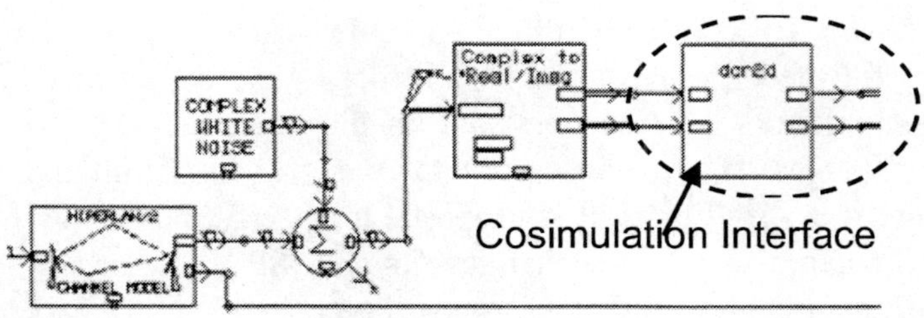

Figure 12-7. Generated interface model in the SPW schematic (detail)

The Verilog-AMS netlist is compiled. Based on this, an SPW symbol can be created in an SPW library. It is supported by the SPW block wizard. The symbol can be easily placed in the SPW schematic of the WLAN testbench as depicted in Figure 12-7.

The cosimulation is started from the SPW simulation manager. SPW-AMS simulator is chosen as the simulation engine. The AMS Designer can be executed in the background (batch mode) or interactively (GUI) with full debugging and visualization features for the RF subsystem.

Some extensions to the existing models may be necessary to achieve the desired simulation results. The RF subsystem model can add delay and phase shift to the signal path. If the digital baseband model of the receiver does not contain synchronization and phase recovery algorithms, the evaluation of

BER (bit error rate), PER (packet error rate) and constellation diagrams can fail. The user should pay attention to the following points:

- *Signal level adaptation*: In system level simulation the signal power is often normalized. Signal power adapters are inserted to provide appropriate signal levels for the analog blocks.

- *Scheduling of simulation*: Defined sampling rates are used in digital baseband processing, while the analog part is simulated with continuous time. Correct scheduling of the simulators must be ensured. Resampling is necessary in some cases.

- *Signal delay adjustment*: The system level model often uses a hard synchronization of DSP algorithms to the signal flow. For example it is assumed that the first received bit is also the first bit of a signal frame. If the analog subsystem introduces additional delay in the signal path the algorithms have to be synchronized by a proper delay setup.

The output of the demonstration systems is a BER statistic written into an ASCII file. The BER data is only available when the simulation has successfully finished. Probes must be added to observe some waveforms during the simulation. This can be done in the schematic entry or in the simulation manager. If it is necessary to add probes within hierarchical models of the *wlan_lib* library it is proposed to specify the probes from the simulation manager to avoid changes in the SPW libraries. Another advantage of this method is that the probes can be enabled or disabled in the "Simulation Manager" window.

The signal calculator Sigcalc can be started during a running simulation. It may take some time before Sigcalc has received enough data, then the selected signals are displayed. The correctness of delay correction and input and output scale parameters can be checked from the *RF_in* and *RF_out* waveforms. The *scat* function can be used with default parameters to see the scatter plot of the quadrature amplitude modulation (QAM) signal. The channel model was set to "AWGN" instead of "Fading" in the top level schematic of the IEEE demo. Otherwise the effects from fading will hide the effects from the nonlinearity of the RF subsystem. Two different configurations of the cosimulation are shown in Table 12-1.

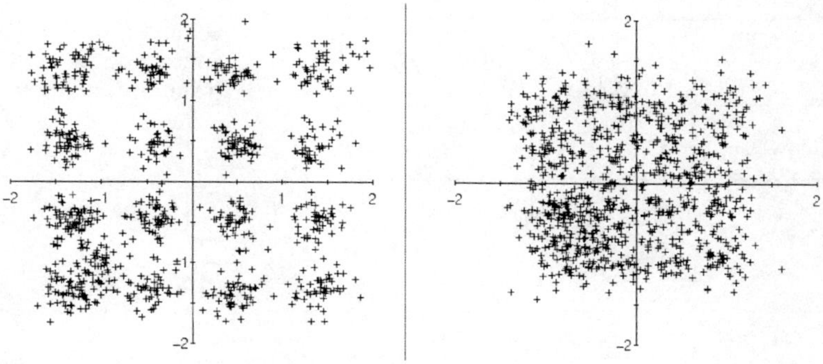

Figure 12-8. Scatter diagrams from IEEE 802.11a demo (left configuration: low input level, right configuration: high input level)

Table 12-1. Simulation of different system configurations

Configuration	Low Input Level	High Input Level
Input scale	0.1	0.2
Output scale	0.008	0.004
Compression Point	10.0	10.0
Result (100 OFDM blocks)	RF out is compressed partially, scatter diagram with few deviations, BER is 0	RF out is compressed partially, worse scatter diagram, BER is still 0

The result plots in Figure 12-8 depict the scatter diagrams. A significant signal deviation can be at high input levels. While the scatter plot is strongly disturbed the digital error correction algorithms can still achieve a bit error rate (BER) of zero. Only further increasing of the input scale causes bit errors. A BER of $5.6 \cdot 10^{-4}$ was measured in the simulation for an input scale of 0.3.

This example demonstrates how the impact of a nonlinear RF subsystem could be considered in system simulation. The BER measurement shows the impact of the RF subsystem on the system performance.

12.4 Generated Black Box Models

The cosimulation method allows very accurate analog and RF subsystem modeling up to transistor level accuracy. However in some cases the simulation performance is not sufficient for extensive system analyses like BER evaluation. Black box behavioral models can be used to accelerate the simulation. The principle of table based models is depicted in Figure 12-9.

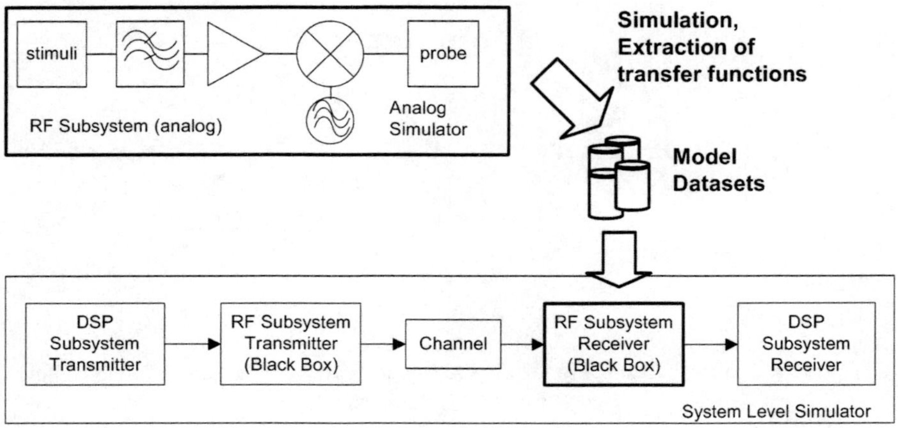

Figure 12-9. Principle of black box modeling

Generated black box models are suited for bottom-up verification, since these models are extracted from detailed models at block or transistor level. They represent a component or a subsystem without information about its structure. This can be realized by using data sets with measured characteristics.

In a design flow for communication systems this modeling technique is helpful to bring the behavior of analog system parts into the system level simulation of the full system. Analog or RF simulators are used to extract the transfer functions of the RF front-end. They provide high simulation accuracy. Non-ideal effects between components (for example impedance mismatch) are considered. RF simulation technology as outlined in Section 3.2 is used to analyze and characterize the RF subsystem. The extraction of the model dataset can be automated.

Corresponding models in the system simulator represent the behavior of the analog front-end based on the generated data files. A great benefit of this technology is the automated generation of complex baseband behavioral models (Section 4.3) for RF front-ends from passband behavioral or transistor level models, which provide high simulation performance.

The end-to-end transmission system can be tested with signal distortions that originate from the RF subsystem. Now it is possible to adjust and optimize the DSP part in a number of simulations. Since the analog part is a black box model, no optimization of the RF subsystem is feasible in the system level simulator. New model data sets must be extracted if the analog part was changed.

Good model accuracy and improved simulation performance are the main advantages of these models. An extracted model can be used many times.

The analog simulator is only used for model extraction. This reduces licensing costs. The IP of the underlying circuit is protected due to the black box architecture. So the model can be provided to third-party system designers.

Characterization and generation of the model data sets can be very time consuming. This is disadvantageous because the model must be extracted after each change of the RF subsystem. Since the RF subsystem is modeled as black box no optimization of the RF part is feasible in system level simulation. It restricts the application of this modeling technology to bottom-up system verification.

Black box models can represent single building blocks of the RF front-end like mixer, filter and amplifier (for example low pass equivalent models [JBS00]) as well as complete RF subsystems [MoC98].

The J&K model approach

J-models are designed for modeling complete transmitter front-ends. K-models represent receiver front-ends. This means that the transmitter subsystem has a complex baseband signal input and a passband (carrier frequency) output. The receiver subsystem has a passband (carrier frequency) signal input and a complex baseband signal output. The model extraction is based on SpectreRF simulations. The generated models can be used within the system level simulator SPW.

If J&K models should be extracted for separate parts of the RF subsystem it is necessary to add ideal mixers to ensure that the type of input or output signal is correct. After extraction (on the SPW side) the models have complex baseband input and output ports.

The J-model can be used to evaluate adjacent channel power ratio (ACPR) and error vector magnitude (EVM) of transmitters at system level comprising intermodulation, spectral regrowth and harmonic distortion. The K-model provides end-to-end bit error rate (BER) evaluation for receivers. To evaluate the impairment of a blocker it is necessary to model the blocker in SpectreRF for it to be present during model extraction.

K-model example

The direct conversion receiver model which was designed for the cosimulation was also used for the K-model extraction. Some modifications are necessary to use the model for K-model extraction. An ideal mixer was added in front of the receiver to convert the input passband signal immediately down to the complex baseband representation. This is because the receiver front-end is described completely with complex baseband behavioral models. Input and output jig were added. These are specialized port models, which are used to determine input and output ports of the RF front-end. The resulting schematic is shown in the upper part of Figure 12-10.

The analog design environment ADE is started from the schematic window. Some simulation parameters (for example simulation sweep ranges) must be specified in the ADE window. Then an OCEAN script is generated, which contains all analysis commands for the model extraction.

The script is executed in a shell. It starts a set of simulations and manages the generation of the K-model dataset. SpectreRF is used as the simulation engine. Periodic steady state analyses are used to extract the transfer function of the receiver.

The model extraction can take some minutes or up to several hours depending on the complexity of the receiver model. During this time the SPW schematic can be modified. The nonlinear K-model template is placed from the *SpectreRF-SPW-Link* library into the schematic (bottom of Figure 12-10). It is configured by specifying the directory that contains the extracted data. The full system model is now ready for simulation.

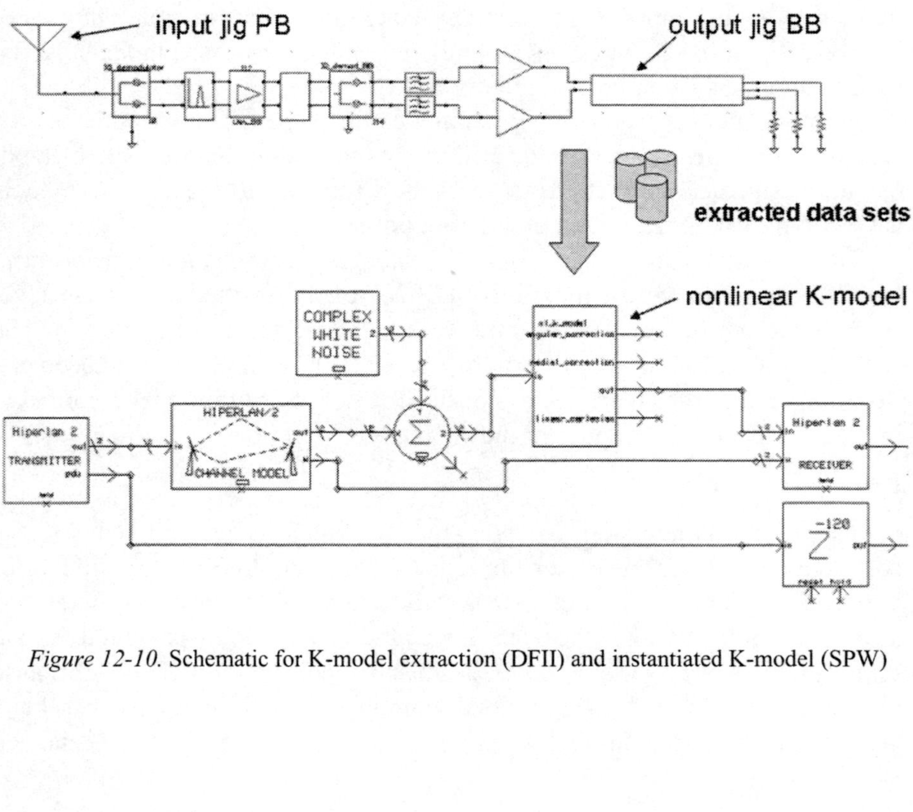

Figure 12-10. Schematic for K-model extraction (DFII) and instantiated K-model (SPW)

The K-model was used within a QAM16 transmission testbench. The simulation result (Figure 12-11) shows the effects of compression and AM/PM conversion of the QAM symbols.

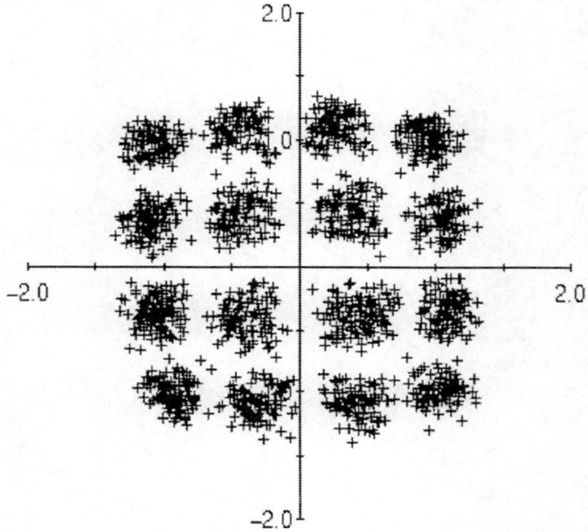

Figure 12-11. Scatter plot from QAM 16 testbench with K-model

References

[Acc04] *Verilog-AMS Language Reference Manual*, Version 2.2, Accellera, 2004.

[APT03] Ashenden, P.J.; Peterson, G.D.; Teegarden, D.A.: *The Designer's Guide to VHDL-AMS*. Morgan Kaufmann Publishers, 2003.

[Ash02] Ashenden, P.J.: *The Designer's Guide to VHDL*. Morgan Kaufmann Publishers, 2nd edition, 2002.

[BCP74] Boyle, G.R.; Cohn, B.M.; Pederson, D.O.; Solomon, J.E.: *Macromodeling of integrated circuit operational amplifiers*. IEEE J. Solid-State Circuits SC-9 (1974) 6, 353 - 363

[Bes03] Best, R.: *Phase-locked loops: design, simulation, and applications*. 5th edition, McGraw-Hill, 2003

[BLM04] Barry, J.R.; Lee, E.A.; Messerschmidt, D.G.: *Digital Communication*. 3rd edition, Kluwer Academic Publishers, 2004.

[Bra67] Branin Jr., F.H.: *Transient Analysis of Lossless Transmission Lines*. Proc. of the IEEE 55(1967), 2012-2013.

[Cad03a] *SpectreRF User Guide*, Product Version 5.0, Cadence Design Systems, 2003.

[Cad03b] *OCEAN Reference*, Product Version 5.0, Cadence Design Systems, 2003.

[Cad03c] *Cadence Verilog-A Language Reference*, Product Version 5.0, Cadence Design Systems, 2003.

[CaS91] Casinovi, G.; Sangiovanni-Vincentelli, A.: *A Macromodeling Algorithm for Analog Circuits*. IEEE Trans. on CAD 1991, vol.10. no.2, pp. 150-160.

[ChB99] Christen, E.; Bakalar, K.: *VHDL-AMS – A Hardware Description Language for Analog and Mixed-Signal Applications*. IEEE Trans. on Circuits and Systems-II 46 (1999) 10, pp. 1263-1272.

[ChD87] Chua, L.O.; Desoer, C.A.: *Linear and Nonlinear Circuits*. McGraw-Hill, 1987.

[CoC92] Connelly, J.A.; Choi, P.: *Macromodeling with SPICE*. Prentice Hall, New Jersey, 1992

[DeK69] Desoer, C.A.; Kuh, E.S.: *Basic Circuit Theory*. McGraw-Hill, 1969.

[JBS00] Jeruchim, M.C.; Balaban, P.; Shanmugan, K.S.: *Simulation of Communication Systems*, Second Edition, Kluwer Academic Publishers, 2000.

[Kam92] Kammeyer, K.D.: *Nachrichtenübertragung*. Stuttgart: B.G. Teubner, 1992.

[Kun02] Kundert, K.S.: *Accurate and Rapid Measurement of IP2 and IP3*. Version 1b, www.designers-guide.org, 2002

[Kun95] Kundert, K.S.: *The Designer's Guide to Spice and Spectre*. Hardcover 1st edition, May 1995.

[Lee01] Lee, D.C.: *Modeling Timing Jitter in Oscillators*, White Paper, Mentor Graphics Corporation, June 2001

[LiZ97] Litovski, V.; Zwolinski, M.: *VLSI Circuit Simulation and Optimization*. London: Chapman & Hall, 1997.

[LuB00] Ludwig, R.; Bretchko, P.: *RF Circuit Design – Theory and Applications*. Upper Saddle River: Prentice-Hall Inc., 2000.

[MGC04] *ADVance MS User's Manual*. Mentor Graphics Corporation, 2004.

[MoC98] Moult, L.; Chen, J.E.: *The K-model: RF IC Modeling for Communications System Simulation*, IEE Colloquium on Analog Signal Processing, October 28, 1998.

[Nav93] Navabi, Z.: *VHDL – Analysis and Modeling of Digital Systems*. New York: McGraw-Hill, 1993.

[PeM91] Pederson, D.O.; Mayaram, K.: *Analog Integrated Circuits for Communication – Principles, Simulation and Design*. Kluwer Academic Publishers, 1991.

[QNP93] Quarles, T.; Newton, A.R.; Pederson, D.O.; Sangiovanni-Vincentelli, A.: *SPICE3 Version 3f3 User's Manual*. Berkeley: University of California – Department of Electrical Engineering and Computer Science, 1993.

[Sch90] Schrüfer, E.: *Signalverarbeitung*. München, Wien: Carl Hanser Verlag, 1990.

[Std03] *IEEE P1076.1.1/D1 Draft Standard for Standard VHDL Analog and Mixed-Signal Extensions – Packages for Multiple Energy Domain Support*. December 2003.

[Std99] *IEEE Std. 1076.1-1999: IEEE Standard VHDL Analog and Mixed-Signal Extensions*. Approved 18 March 1999.

[TiS02] Tietze, U.; Schenk, C.: *Halbleiter-Schaltungstechnik*. Heidelberg: Springer-Verlag, 2002.

[Van00] G. Vandersteen et al.: *A methodology for efficient high-level dataflow simulation of mixed-signal front-ends of digital telecom transceivers*, Design Automation Conference (DAC) 2000

[Vla94] Vladimirescu, A.: *The SPICE Book*. J. Wiley & Sons, 1993.

[VlS94] Vlach, J.; Singhal, K.: *Computer Methods for Circuit Analysis and Design*. Van Nostrant Reinhold Co., 2nd edition, 1994.

Index